A Tale of Trust and Hard Knocks

The Biography of F. Donald Caswell

A Tale of Trust and Hard Knocks is an account of the life of F. Donald Caswell. The author has done his best to ensure the accuracy of the account, however, over time memories fade and change. Any differences in the author's account and what others may remember were not intentional or intended to harm others. The author, publisher and all those associated with this book, directly or indirectly, disclaim any liability, damage, loss or injury resulting from this account.

A Tale of Trust and Hard Knocks
Second Edition 2017
Copyright © 2000, 2017
All Rights Reserved to F. Donald Caswell

Published by Morten Moore Publishing, LLC
PO Box 881
Flagstaff, Az 86002

ISBN 978-0-9672576-4-8

No parts of this publication may be reproduced, stored in a retrieval system, or transmitted in any form or by any means, electronic, mechanical, photocopying, recording, or otherwise, without the prior written permission.

Contents

Chapter 1
**A TIME TO BE BORN:
THE DAYS OF MY BIRTH, EARLIEST MEMORIES
AND KIDHOOD**
- 1 -

Chapter 2
AN EDUCATION IN SCHOOL AND HORSES
- 9 -

Chapter 3
COLLEGE BOUND
- 29 -

Chapter 4
NO COINCIDENCE
- 55 -

Chapter 5
**MACUMA, ECUADOR
with the GOSPEL MISSIONARY UNION**
- 79 -

Chapter 6
EVERYDAY LIFE IN ECUADOR
- 109 -

Chapter 7
A TIME TO RETURN HOME
- 195 -

*The secret things
belong to the Lord our God,
but the things revealed
belong to us and to our children forever,
that we may follow
all the words of this law.*
Dueteronomy 29:29

*I planted the seed,
Apollos watered it,
but God made it grow.
So neither he who plants
nor he who waters is anything,
but only God
who makes things grow.*
I Corinthians 3:6-7

Chapter One

A Time to be Born: The Days of My Birth, Earliest Memories and Kidhood

The name and date on the birth certificate read to the effect that I was born on the 19th of June 1929 at the Good
Samaritan Hospital in Phoenix, Arizona. June in Phoenix: it had to be a hot day; I have disliked hot days ever since. I am a cool-weather person, not cold but cool weather. There are limits to what this body can stand, to what it enjoys; the difference between cold and cool is one of them.

That birth certificate also says my given name was to be Floyd Donald. Curiously enough, I was commonly called Donnie. I think that name fits me better than Floyd. The way I figure it is Mom and Dad knew they would have to yell and holler to get my attention. So they stood at the back door and practiced a time or two. 'Donnnnnnnie' was easier to yell than 'Flooooyyyyd.' It was easy to see that 'Donnie' would just suit their purposes better.

By the time, my brother, Derral, came along Mom and Dad knew more about this naming process. The name 'Derral' is good for common conversation but when they had to yell to get his attention they belted

1

out, 'Derral Dean.' If you give it a little try, you quickly realize those two names go well together for yelling.

But it was my understanding until age 21 or so, that my given name was Donald Floyd. I learned the truth at age 21 when I first saw my birth certificate.

Having introduced myself as Floyd Donald, I should introduce the rest of my family as well. Derral and I were the sons of Layne and Opal Caswell. My mother's parents, Ida Bell and Pink Moore lived near us through much of my childhood and had a great influence on my life. In my teen years, my parents added my two sisters, Lavonne and Linda, to our family.

The earliest memory I have is of a goat being kept in our backyard. During those first couple of years of my life, I had stomach problems and was not strong so the doctor prescribed goat's milk for me.

A second memory is that Pop and Grandma kept a milk cow, which provided for all of us. My job was to go through the backyard gate to their house and bring back the pail of milk. On alternate evenings Derral made the trip. They always filled the pail right to the rim and we had to be so careful as to not spill it. I always wondered why they filled it so full. Not only was it always full to the rim, the pail was about as wide as it was deep so it was easy to spill. Several quarts of milk in a pail is pretty heavy for a four-year old boy.

My parents were not wealthy but they were reasonably content with what they had. We always had enough to eat, a comfortable house, sufficient health care and clean clothes.

We had lots of good times during our childhood: family picnics in the park, evenings with friends, visits by relatives from far and wide and a vacation in a National Park every year or so. There were lots of practical jokes on one another, a chase through and around the house, across the yard over to grandpa's house. Mother and Dad doing the dishes got into dishtowel and dishrag snapping contests with Derral and I urging on our particular hero of the moment. I remember helping with preparing vegetables for canning, making laundry soap from ashes, animal fat and lye.

We were properly disciplined. I don't know that the discipline dealt us would be acceptable under today's liberal psychology. Never-the-less it worked and I don't think I was damaged emotionally by whatever was dealt us in the name of discipline.

Pop and Grandma lived next door. Dad and Pop, between the two of them, owned the two lots our houses were built on as well as the rest of the land in the corner of the two roads. We used the land for fruit trees, a large vegetable garden and a playground. The plot of land across the road from our homes lay fallow. Derral and I used it as the site for whatever adventures we might consider.

I started my education in September of 1935 at the neighborhood school just a half mile from my home, so close that I walked home for lunch. This was not always what I desired. Mother usually made Derral and I rest a bit before returning to school. For her, resting meant we were to *lay down on your bed*. I figured I missed out on lots of fun during the lunch recess so I sneaked off, returning to school. She was rather upset when I came home from school that afternoon. One of those times when I learned of discipline!

My first year in school was not a good one. My teacher instilled only fear of authority. She was very stern, a very poor disciplinarian. There was one overpowering incident I shall never forget. There was one boy in the class who was kind of odd. One afternoon the teacher told him to stay after school due to his behavior. For some reason I had stayed as well. When I started to leave he took the opportunity to make a run for the door. The teacher screamed at me to stop him. Forcing him into a chair, she began to wrap a very long piece of cotton rope around and round him and the chair. I distinctly remember thinking that she certainly did not know how to tie a person to their chair.

A year or so earlier Derral and I had decided we were going to milk the family cow. We had tied the calf to a post in the corral, the cow to another and before we got to the milking our grandmother caught us. We had tied the two animals pretty well. Due to that earlier experience I figured I knew how to handle a rope.

I enjoyed the years in the second, third and fourth grades, redeeming my belief in good education. My social life was made up of friendships at school, kids in the neighborhood and from our attendance at Hope Chapel. While I was in the second or third grade, Dad and Mother bought me a two-wheel bicycle; Derral was still in the tricycle stage. The bicycle was a little different as the wheels were small, maybe twelve inches in diameter. I rode it to school and back most every day.

One day while on my noontime transit something just went wrong. I was approaching the bottom of the hill, pedaling hard, picking up speed to make it up the hill. Somehow my foot slipped off the pedal and I fell, splattered out all over the very rough pavement. I had worn short pants that day and my knees were severely scraped, the blood flowing freely. I got up and went back home for emergency medical services. I didn't make it to school that afternoon.

We moved south to Montecito in the summer of 1939, where I attended the fifth and sixth grades. As Derral and I had not grown up in this community, we both experienced a good deal of harassment by older and bigger boys. I had a fight or two there, both were over someone else picking on Derral.

One incident in particular remains with me. The school playground was built on three terraces. One afternoon while playing ball on the upper terrace, I noticed Derral was being knocked around on the bottom terrace. I took off running, arriving on the run and swinging at whoever it was knocking him around. Between the surprise of my unexpected arrival and the accuracy of my fists, the fight was quickly won. No one ever bothered him again at that school.

In Montecito, we lived on the Cudahy Estate about three miles from East Beach. Almost every morning, except Sundays at about six o'clock, we would go down to the beach with two neighbor boys and their dad. We all would swim out to the raft, rest a few minutes and then swim the quarter of a mile down to the next raft and then into the beach. We boys would get tired along the way and the neighbor, Mr. Moffatt, would let one of us at a time hang on to him while we rested. We had a great time. After returning home and washing off with a water hose outside, we dressed and ate breakfast. Then there were chores around the house before we left for school. During the summer, we went off Vacation Bible School in the morning and back to the beach for a couple of hours in the afternoon.

I learned a great many things while living on the Estate. I asked Dad one day where turpentine came from. He told me it was distilled from the resin of pine trees. One Saturday, in search of something to do, I thought about all that resin seeping out of the trees over by the main entrance. I told Derral and the neighbor that I was going to make some turpentine and the project was on.

After getting permission to play in the pines, we got a coffee can and went to collect pine resin. We brought the resin to the laundry room. There was a gas stove, ironing boards attached to the wall, and lots of table space in the laundry room.

I knew that to distill something you had to heat it. I did not know that turpentine was highly flammable. We turned the stove up high, placed our can of resin over the flame and watched the resin begin to melt down. Of course the inevitable happened. As the resin melted it became very volatile and the heavy fumes rose to flow out over the top of the can and then caught fire.

I was scared to death. I could imagine the whole building going up in flames and it was not our building. While the neighbor boys took off for home, Derral stayed with me not knowing what else to do. I turned the stove off. Somehow the can cooled and the turpentine quit burning. This was a memorable lesson. I really did learn a lot while living on the estate.

Mother and Dad spent much time in the evenings with Derral and I reinforcing what little basic education we were getting at school. They read to us and with us. Dad helped as much as he could with arithmetic. We made the drive into town every Sunday morning, attending church at Hope Chapel. We usually attended Wednesday night prayer meeting as well.

During this period of time I accepted Jesus as my savior and Lord. I was twelve or thirteen years old. One Sunday my teacher talked about salvation and at the end of the class she asked me if I would like to accept Jesus into my heart. I knew in my heart I needed to do so and I said, yes, I would. We prayed asking forgiveness for my sin. She urged me to go tell Mother and Dad which I did. That decision would direct the entire course of my life. I have never regretted the decision.

Chapter Two

AN EDUCATION IN SCHOOL
AND HORSES

In 1941, as we entered World War II, my parents moved to Los Angeles. My father moved from one job to another, all involving hard physical labor: in a meat packing house carrying quartered beef carcasses, working in an aluminum foundry, and one with a pipe-lining contractor.

The war brought vast changes to our society, many of those changes met the need of the moment. For many men and women their efforts supported the war even as their motivation was rooted in the pursuit of power and money.

When my mother was a teenager, she worked with her mother at the *walnut house* in Goleta. After she married, she worked at the lemon-packing house in Santa Barbara. Working outside the home was not a new experience for her as it may have been for some women. She found a job midst men and women in the war industry in Mayfield. I believe that this work gave her a self-confidence she didn't have before. The work in the lemon and walnut houses was considered to be only *women's work*, unlike the jobs that men held. Later on that experience gave her confidence to go out and get a job in industry after her children were grown. In those later years, she was quite proud of the fact that her male bosses commended her for the fine work she did in soldering on delicate circuit boards.

Mother was not at home when we left for school nor was she home when we got back home from school. With both mother and dad working, Derral and I learned to make our own breakfast. Unlike the early years, we carried our lunches to school. We became latchkey kids.

The time in the Los Angeles area was a difficult time for all of us, especially for dad. We were issued coupons for sugar, meat, coffee, butter, and gasoline. If the item was important to the military needs, it was rationed. I don't know that anyone went hungry or failed to arrive where they needed to go due to rationing. The use of coupons was a fact of life needed at this time in our nation's history.

To say the least, the adjustment to life in Maywood for Derral and I, was not easy. We grew up in the fields and groves of Santa Barbara and Montecito, knowing nothing of the streets of a busy city. Our family attended a Southern Baptist church two blocks from our house. Dad and Mother attended a fine Bible class of people their age, people with whom they maintained contact for years to come.

Two incidents occurred at the church that influenced my outlook on life. The first dealt with personal security. During a Sunday evening service two junior high age girls left the service to use the restrooms in the area of the building behind the platform. The building had several doors to the outside, none of which were locked. Three minutes after they left, the congregation heard two screams from the rear of the building. Several men jumped up, running through the doors at the front of the sanctuary and out the main entrance around the outside.

The girls told their parents that when they entered the restroom, a man grabbed them. As they screamed, he released them and ran out of the building. He was never apprehended. Being in junior high school, I will never forget the screams.

The second incident involved our joining a club for boys. The church sponsored youth groups using as leaders, young men and women attending university or Bible colleges in the city. With our parents' encouragement, we attended the second meeting and learned a rite of initiation was on the agenda. We had no idea what that might include. There were 30 or 40 boys of all ages present. Some knew the ropes and some didn't have a clue. For a kid who kept to the outer fringes of groups, my fear was terrible.

We were forced to run between two rows of boys who were yelling and screaming, pounding on us, trying to keep us from arriving at the end of the line. Then, we were subjected to severe paddling on our back sides and other ridiculous forms of intimidating violence. I was determined I was not going to yell. Long before the night was over I had decided that I wanted nothing more, ever, to do with that sort of thing. The incident confirmed in my mind that I never wanted to join any kind of club activity, church-related or otherwise. As I detest any sort of hazing, I determined that no sort of membership was worth undergoing such humiliation.

In our free time, away from home and church, both Derral and I had bicycles. We were allowed the freedom to ride in probably a 25 square block area, developing friendships with other kids and adults. Derral and I got the idea of mowing lawns in order to make some money. We borrowed a small trailer to pull behind our bikes from someone in our family. After loading a lawn mower and rakes on the trailer, we made our rounds. The lawn mower was the old reel mower that had to be pushed. Derral was the better salesman so he got the jobs and I did a bit more of the work.

I attended seventh and eighth grades at a two-year junior high school near our home. I did not like the school. I did not like the teachers. I did not like the kids. I must admit though, looking back on the experience, I can think of a number of things I learned in spite of my dislike for school, kids and teachers. We all had been uprooted from our home turf and thrown together, a situation bound to cause all kinds of psychological problems.

Under today's standards, the school was sadly under equipped with little to offer other than classrooms, one wood-shop and a bare dirt playground. The school was even lacking an auditorium for assembly of the students.

I was a small boy during my school days. Being small, I provided sufficient target for the larger boys as they figured I was no danger to them. I was picked on too many times. More than a few times I had to fight my way out—sometimes I was successful, sometimes not. I was quick with fists and feet, so some learned that picking on a small opponent did not always mean a sure and unscathed victory for them. Undoubtedly, my quick temper contributed as much to my troubles as anything else.

School & Horses

Living in Los Angeles, the war was everywhere so to speak. Everyone seemed to have someone in the military fighting somewhere. Everything was done for the war effort. We could not escape it. Factories in Los Angeles were building aircraft of all sorts. In the port cities, ships of every description were being built, all for the war effort. Every industry in the city was working day and night. Life went on twenty-four hours a day, every day of the week.

We saw things we had not seen before. We saw people doing things that we had never seen done before leaving us standing wide-eyed in wonder. We saw that people can and will change in a new environment when no one knows who they are. Yet their character and personality remained the same. Some continue to live by their old principles; some build a new system to accommodate life in their new community. I found it interesting to watch what people did when the principles by which they thought they lived and died were challenged, leaving them free and open for change.

I remember playing in the street and hearing the drone of an airplane in the distance. I looked up to see formations of fighter planes, dive-bombers, heavy bombers and transport planes. Quietly all the games stopped, kids and adults watched in awe at things in the sky they had never seen before. Each week we paused as hundreds of airplanes in formation passed overhead.

When we traveled to Santa Barbara, we watched troop trains filled with soldiers and flatcars loaded with trucks, tanks and other vehicles. We witnessed a hundred and fifty trucks in a convoy moving in the opposite direction we were driving, all loaded with men and materials. They were headed for rail heads to be shipped to the east coast. From the east coast ports they would be loaded on to merchant and troop ships destined for the war in Europe. Other convoys were destined for ports on the West Coast to be shipped to the war in the Pacific.

Anti-aircraft balloons tethered by cables to the ground flew over the city. The balloons were designed to snare attacking aircraft. Sometimes at night we practiced air-raid drills, covering all our windows with blankets to keep all light from getting out of the house. Men and women were assigned to patrol the neighborhoods to ensure that we all were in compliance. The penalty for non-compliance could result in jail-time. Stepping outside, we watched powerful search lights scanning the sky for foreign attackers and when the lights locked on to an airplane a shout of awe went up from the city. Was it Japanese? German? What is it? The next day at school we discussed the events of the night past. The rumors of what we had seen were made into certainty before an hour went by. The time of war is a time of rumor.

I remember listening to the eyewitness reports of men who had the good fortune to receive a leave or furlough from the war-front to come home. Most were wounded to received the leave. One Marine, a friend of my parents told us of fighting on the islands in the Pacific. He described the Japanese as so strapped for metal from which to make bullets that they used wooden bullets. Wooden bullets also killed and wounded men. I wondered at each tale as I sat and listened.

After two years in Maywood, my dad was offered a job in Santa Barbara. I have forgotten the details of the move, but we were all so glad to be moving back home.

Pop brought his horse trailer down. With other vehicles, we made the move in a caravan up the coast. In 1942, the government began rationing certain commodities such as gasoline, sugar and meat. Yet, dad and Pop obtained the required gasoline.

Our return to Santa Barbara brought lots of changes to our lifestyle. The economy was hot. People were earning more money than they ever had and as the need for war materials diminished, people began to spend more on themselves. We were not much different from anyone else.

After our return to Santa Barbara, my parents decided their family could grow by one more child. My sister, Lavonne, was born when I was fourteen. They had agreed on one child and my mother now had a girl to raise in ribbons and bows. Much to her chagrin, my second sister, Linda, arrived two years later. Linda would be well loved, living near them and a comfort in their old age.

My high school days were full of horses, swimming, running, girlfriends, and getting together with other young people at Hope Chapel. I chose my friends from either the church youth group or those who were from Goleta. I don't really remember much about the matter of school classes, homework and grades. I think I was too much interested in taking life day by day and letting the future take care of itself.

As part of my class load, I enjoyed courses in advanced mathematics. Geometry was interesting but it seemed that I was always on the edge of understanding a concept when it was time to move on. I really enjoyed my trigonometry class and saw lots of usefulness in it. Choosing to take physics rather than chemistry was something of a disaster—I fought to get an average grade from a teacher that neither I nor many others could respect. I enjoyed English literature and history classes. I stayed with the college prep courses, refusing the shop classes. I liked school but it was hard for me.

Of my extended family, none are sports-minded people as so many are today. My father enjoyed listening to boxing matches on the radio in the 1930's. During his later years, he watched baseball on television. He had played softball in school when he was a teenager and my parents took us to the girls' softball games. I also enjoyed the *donkey baseball* games of the 1930's. My parents did teach Derral and I to play tennis; that, along with my interest in horses, gave us good dating activities. I tell you all this to indicate that I did not come from a family that spent time in athletic pursuits.

I was required to take Physical education classes in high school where we spent time running races of various lengths. I had won all the short distance races with great ease. After the class was over the coach approached me and asked why I had not signed up for track. I told him that I was not fast enough and so it had not entered my mind to do so. He said that after watching the period's activities it was evident to him that I indeed was qualified and he would like me to give it a try.

This was something that had never happened in my life: a coach or teacher telling me I was qualified to enter some such activity and then encouraging me to do so. Coach Clarence Bishop became a friend. I did try out for the track team and due to my small size made the 'B' level. I ran the 100-yard and 220-yard sprint races and the four-man 440-yard relay races. I set several records that stood for two or three years and gave some proud moments to our family in Santa Barbara. I made the team both my junior and senior years.

Along with running in high school, I began to swim more. I now laugh at the incident, probably in the eighth grade, that motivated me to learn to swim well. Derral, our neighbor and I were swimming in the Santa Ynez River one warm summer day. We had arrived at a swimming hole we had never used before. Two girls a couple of years older than me came down and dove into the water, swimming as if they were fish with such speed and grace as I had never ever seen before or even thought possible, especially for a girl. At that moment I decided that if a girl could swim like that, then I must learn to do better than them. To know they were so much better than me hurt my ego but motivated me tremendously. Up to that point in my life I had been content to just dog paddle along. From that moment I began to learn to swim. I watched others, I asked questions. I learned to keep in tune with what the motions of my arms and legs were doing. It felt good to swim well, fast and easy. I just kept on learning. Thank you, ladies!

In both junior high and high school the young people we knew were divided into two groups: The farm kids from Goleta and the city kids of Santa Barbara. I belonged to the farm kids from Goleta, though I had not grown up with them. We lived between the two communities which were separated by open country, unlike today.

Pop and Grandma had sufficient land to be able to

build a small barn for a couple of horse stalls and hay storage. The field across from our home was available for pasture. Pop wanted a horse around the place as some people want a dog. He had a dog. Now he wanted a horse to ride. A horse represented the *good old days*, though he vigorously denied that there was anything good about the *good old days*. Weekend or the odd off-day, we could find him fussing around with his one or two horses. Riding was enjoyable, the horse more than a mere companion. His deteriorating left leg made riding something of a chore.

Derral and I were into our teens and I suppose that both Dad and Pop felt that we needed something to do to keep us out of trouble. Together they purchased horses and feed, giving Derral and I the responsibility to care for them. Horses served many purposes in our lives. Unlike Derral, my inclination toward horses gave Pop an outlet for his desire to teach a boy the art of horsemanship. From age 14, there were horses in my life until I left California.

Dutch was the first horse I learned to ride. She was one of the most amazing horses I have ever known. She was a light gray, stocky horse, weighing about a thousand pounds, standing fourteen hands high. Her gentleness allowed most anyone to do anything around her. She was very gentle, cognizant of what was going on around her. If a child fell off her while she was in motion, she would stop and stand still until the all-clear signal. She was comfortable around other animals until they showed some sign of aggressiveness toward her.

As a trained cow-horse, when she was out on the range, her total attention was on the cattle with which she worked. She assumed her rider could stay on her regardless of how she moved to accomplish the job she perceived was desired of her. Her rider had better be a good one. Pop loved that horse and was very dedicated to her well-being. He was quick to sing her praises. We used her to train younger horses with Pop riding Dutch while I rode the colt. We had some great times in such concert. When we transported her and another horse over the mountains to the Santa Ynez River, we rode and played in the river with her. One year, if the river had been deeper, she would have had to swim. She was probably some 25 years old when she died.

Along with Dutch, we needed a second horse so that Derral and I, or Pop and I, could both ride at the same time. Pop found a four-year-old colt on a ranch in the Santa Ynez Valley. The horse was a gelding and had spent most of his four years out in the hills and mountains. His bloodline was prominently Morgan. Broken only to the halter, he was pretty rough looking but bright and responsive to handling. Dad bought the horse, Pop and I broke him. He turned out to have great curiosity and was very playful but steady. In many respects, he was just a young teenage kid. Dutch and Morocco, as we called him, were quite a pair. Dutch being old and more serious, brooked no nonsense from Morocco. The young horse seemed always to be bugging Dutch and received a few hard bites or kicks from her for his efforts.

One of the most difficult horses I trained was a tall horse owned by Joe Dal Posso the owner of a tire company. Pop shod Mr. Dal Posso's horses so he knew about the horse. Someone else had originally broken the horse and Mr. Dal Posso had just let the horse go without riding him. He forgot all he had learned and then gained the mean habit of turning his tail to kick when one tried to catch him to put a rope on his halter. Joe told me he would give me $75 to ride and train him again. Before I was through with the horse, I thought he would break me. I earned every cent he paid me riding before daylight and then another hour with the horse after school. Late one afternoon out in the field, the horse got mad, reared up and fell over backwards. The cantle of the saddle caught the bottom of my pelvis pining me to the ground for a moment. I thought my pelvis was cracked, my back was broken or something. It hurt for many weeks but I didn't dare tell my parents as they would have made me quit riding. Earlier in the year the doctor had prescribed a brace to cure and strengthen a lower back problem. To have all that expense negated by a half-broken horse would have been beyond reason. It was a happy day when I turned the horse back to Mr. Dal Posso and was handed a check for $75. He was a well-broken horse when I finished.

Pop tried to start me on the road to becoming a horse trader in the strict sense of that phrase. Despite his best effort, I was not inclined in that direction. The incident was profitable to me financially. He profited in the fun he reaped from it.

Someone had given him a jack donkey for the price of hauling the animal away. The animal was worthless for anything except breeding. I am sure Pop had plans that neither his friend nor I knew about. He never expended that much time and energy on something

so apparently worthless without reason. Pop's ear was always open for information he could use. He certainly had no use for a jack as breeding stock. He brought the animal home and tied him in the
pasture. I, of course, was very curious as to why he wanted the animal and what he planned to do with it. He said the male donkey was mine if I wanted it.

A boy of sixteen like me had no idea what to do with such an animal. But Pop was way ahead of me. He suggested I trade it for a horse. This was a good idea but who would be stupid enough to trade this animal for a horse? A few days later, I came home from school and found a horse tied in the pasture—in place of the donkey. Hurrying over to Pop's house, I learned he had found a man who wanted the donkey and paid $25 for him. Pop then said he knew a man who had a pretty good horse for sale for $50. He suggested that since he had the trailer out on the road delivering the donkey, he might as well go out and pick up the horse. I owed him $25. So far, so good. The new horse, about six years old, was well broken. A good horse for $50.—there was something not quite right here. Pop and I wondered but that was part of the fun of horse-trading.

Some days later, on a beautiful summer evening, Pop and I were riding down Hope Avenue. As we rode we talked about what could be wrong with this apparently good horse. Riding next to a lemon grove, my new horse just suddenly broke in two, as the expression goes, and headed out through the grove. He was a heavy, strong horse, not at all mean. He just wanted to go his own way and his way was back to the barn, however and as soon as he wanted. I had put on my favorite spurs, which were small and lightweight with very sharp rowels. When the horse made his first jump, I jammed my spurs into his sides and pulled his head around into the opposite direction he wanted to go. Eventually we got back to Hope Avenue where Pop was waiting. As I rode up to him, he was sitting on his horse laughing.

"Well, that's the first time I ever saw daylight between you and a saddle," he commented. He laughed some more. "Now we know why that guy wanted to sell such a good horse for so little money." He shared my thoughts exactly.

The horse had gotten into the habit of grabbing the bit in his teeth and going where he wanted, usually back to the barn. No one had taken the time to break the horse of that habit. We got home and as I unsaddled and wiped the horse down I saw the damage a pair of sharp-rowelled spurs can do. His sides were bleeding where I had hit him. I never had another incident with that horse. He just needed to know the rider was boss. I kept the horse for another month, training him well. He became a very dependable and well-mannered horse.

One afternoon Pop told me he had found a four-year old unbroken colt for sale. He offered to trade the horse for the colt.

"Sure, why not!" I replied.

Pop traded my horse and $50. for the colt. As with all good horsemen, Pop could see through the good or the bad horse underneath the animal's appearance. The colt was a pretty horse, his coloring white, black and a light brown with black hooves. Best of all, he had a natural pacing gait. Dad was not at all happy but the trade had been made. I spent six months breaking and training the colt. He became a good ranch roping horse.

Of that training process I remember one incident

well. On a Saturday afternoon I was riding the colt out in the field. Suddenly he took his head and bolted for the barn. I had been working on this bothersome habit. As he turned for the barn, he came to the road pavement at an angle. When he turned on the pavement, all four feet slipped out from under him and he landed on his side. I hit the far shoulder of the road, which was rough gravel, landing on the backside of my shoulder and slid for three or four feet. I was wearing a light T-shirt. You can imagine what my back looked like and how I felt. I was a bloody mess, to say the least. The colt got up and ran on into the barn. I got up and went in, caught and tied the horse, all with an intense desire to kill him right then and there. I hurried to our house, knowing I needed help with my back.

Mother was well along in her pregnancy, carrying the child who was to become my sister, Lavonne. She was in no mood to be compassionate with a *crazy kid* like me but I asked for her help with my back. She took me into my bedroom where I lay on my stomach. After picking gravel and bits of t-shirt out of my badly scrapped back she slapped a handful of Noxema cream on me. When the Noxema hit my back I levitated off the bed, yelling, "Ow Mom, that hurts!"

I can hear her response even today, in a very disgusted voice.

"Shut up and lay down. It isn't my back and it doesn't hurt me a bit!"

She continued rubbing Noxema into the wound. She had had enough of her crazy son and his crazy horses. Being seven or so months pregnant didn't help the situation and I didn't really blame her much. I knew she was tired those days. She took such good care of her family and without complaint.

In May that year the superintendent of the San Julian Ranch asked Pop if he would like to come up and help with their branding season. Pop wouldn't have missed it and asked if I could come along. I took the colt on the outing. He pulled the same stunt again but this time we were out in the open country. I just let him run until he stopped and then I ran him some more. Eventually I broke him of the habit.

The next year I sold the horse for about $250. Dad rode him once before we hauled him away. His comment was that the horse was one of the best I had ever had. He really liked the smooth pacing gait in which the horse traveled. I never have been much of a salesman and horse-trading just didn't take with me. I enjoyed breaking and training horses much more than trading for them.

As to being a salesman, it seems to me there must be a particular bent of mind in all good sales people. That mind set disallows the consideration of whether or not the customer really needs the item. I could easily convince myself that the customer might be harmed by buying what I have for sale. I did not want the customer to think badly of me if, after buying the thing, they might find it is unusable to them. What if the client were to realize he had been unwise spending money he really needed in another part of his budget. I am not a salesman. Horse traders have to be salesmen.

I have never had any problem getting up early in the morning. Most times I would lay there reading or go out to clean stalls, feeding and leading the horses out to pasture. During the summer months I would work for an hour before breakfast with a horse I was training, enjoying the most beautiful part of the day.

I had agreed to break and train a colt owned by a man who lived up the road from us. The colt was half Arabian and half Thoroughbred. With the blood line, we would call the animal a rather hot-blooded excitable horse. She was a fine-boned filly and seemed a bit fragile to me. We agreed on a hundred and twenty-five dollars to do the job and brought her down to our barn so as to have her more accessible.

I knew my trouble had just started when I was riding her near the entrance to Hope Ranch one late afternoon. The road ran next to the railroad with a barbed wire fence between the road and the railroad. Unfortunately a steam engine pulling a string of boxcars roared by with all the noise and clamor a train can make. The horse went crazy, bucking, jumping around, whirling one way then the other, trying to run away. I headed the horse toward Hope Ranch intending to get away from the noise and give the horse the idea that she was getting away from the noise and fright. After calming her sufficiently, I returned home. From that time on, every time I passed that spot I made sure I wasted no time getting some distance between us and the railroad. That incident was only the first of several to come.

I got up early one Saturday morning knowing that I had time to ride up into the mountains. The trail I followed leads across Foothill Road and up a quiet and serene canyon. As we climbed, I found the landowner had put a fence across a narrow steep part with a barbed wire gate to pass through before we could proceed up the canyon. I got off the horse opened the gate, walked the horse through, closed the gate and climbed back on.

Before I hardly got seated something scared the colt and she just blew-up, bucking and jumping until she caught her hind foot into the wire of the gate. With the horse tangled in the barbed wire I knew my best action was to step off to try to calm her. I landed on my shoulder and the side of my head knocking me unconscious. I do not remember what happened from that point until I caught her again in a corner of the fence down along Foothill Road. I don't remember running from the gate to where I found her a half-mile away. Somehow she knocked the gate down and ran down the canyon.

As I approached her, her eyes were terror stricken. As I looked her over, I saw her legs and belly were cut and badly scratched. Then, I saw a huge gash in her chest running down between her front legs. I figured she was too badly hurt to ride so I walked and ran two miles, leading her back home. I was wearing high-heeled boots.

I arrived at the barn, tied her up and woke Pop to ask for his help. I was scared to death of the damage I thought I had done to the horse. I knew we would have to call the veterinarian. Pop dressed and went out to see the damage. He called the veterinarian and the owner. The doctor arrived an hour later and sewed the gash, applied sulfa powder and said he would be back on Monday to see how things were getting along. To say the least, that was the end of my horse breaking until the filly got well. Mr. Clausen was upset but very kind and understanding.

Pop was the farrier for many ranches in Santa Barbara County. As a farrier, he was invited most years to participate in the branding seasons of several spreads. Pop worked both the Hollister Ranch north of Gaviota and the San Julian on Highway 1. I got out of school a number of times to accompany him on those round-ups.

For the sheer joy of the lifestyle I wanted to live in the company of my grandfather, nothing could compare with those times. I rode the horse I was breaking at the time. Up before dawn, eating breakfast and then riding across the canyons and hillsides, gathering the cattle to a central location. There the cattle were separated; bulls were taken out of the herd; calves from the cows; barren cows removed to be sold. In the afternoon the bull calves were castrated, all the calves were branded, vaccinated and checked for general health. This activity went on for some four or five days on the Hollister Ranch. The San Julian ran a smaller herd of cattle and the branding season lasted only three days or so.

I have always considered myself fortunate to have done these things as this was the end of an era among the California coast ranches. The several years on the San Julian were of particular interest to me. I particularly remember two of the men who worked on the San Julian. An old Chinese man was the ranch cook and had been for many years. He didn't speak much English. His food was excellent. The top cowboy was a small stocky Indian of Chumash descent, called Chapo or Chico: an ugly little man of magnificent horsemanship ability. He made reatas, reins and headstalls of braided rawhide. He also had two Australian Sheep dogs he had trained to help him work the cattle. One of the two dogs had saved his life in fighting off an enraged bull that caught Chapo off his horse in thick brush.

After I got into college, I worked for a couple of summers on the Dos Pueblos Ranch just up the coast from Goleta. One summer I agreed to work with the horses and cattle on the San Lucas in the Santa Ynez Valley. The foreman put me to work hoeing weeds in the corn crop and I quit after two weeks.

As I read some of the popular literature on men and women who use the gentler methods of breaking and training horses today, I wish I had had their knowledge when I was working with horses. With that understanding, training horses would have been a much more serious endeavor.

Chapter Three

COLLEGE BOUND

The time came to make some serious decisions about college. Pop really hoped I would somehow be able to work with horses and cattle or manage a ranch at the very least. He wanted me to seek a college education to that end. Dad and mother wanted me in business. I understood this to mean owning or managing a shoe store, hardware store, the business of selling something to the public. This just wasn't for me.

I think maybe I was looking for a way of life; from the outside, agriculture and livestock ranching looked like the way of life I desired. Working with horses and cattle was not a good way to make a living at that point in life but as a young man making a living just didn't seem to have any relevance to what I loved.

I saw my grandfather's bias toward the life of horses and ranching as being more open to the realities of his heart. Unfortunately, I perceived dad's bias as 'money is the key to a decent life.' Both men knew poverty intimately, they had spent a good bit of their lives in it. Poverty produced different attitudes in each of them.

Pop's wants and desires were rather simple, more of a freedom from want. Dad seemed to fear poverty greatly. Mother feared it more than dad. Poverty is something to fear, no getting around that.

As I approached my senior year, I considered two colleges: Cal Poly at San Luis Obispo or University of California at Davis. Both schools had extensive agricultural departments which appealed to my interests in livestock. Dad tried to dissuade me from agriculture.

My grade-point level in high school and insufficient financial resources left me with the choice of going to Cal Poly. The program at Cal Poly directs students to work with a production project of their choice. I was able to enter Animal Husbandry in the last months of my first year.

One of the better features that Cal Poly incorporates into their program, for both agricultural and engineering majors, is the use of hands-on individual projects. The project involves the application of what one has learned and is learning in one's major field. I was involved in three projects over the four years I was at Cal Poly: one with poultry, one of fattening pigs for market and one of raising steers for show and market. The last of the three was the most profitable, it also was one in which I shared work and profit with six other students: one of our steers was judged Grand Champion at the Great Western Livestock show in Los Angeles that year. The steer was auctioned off for $10 a pound at the end of the show.

Toward the end of my second year and certainly by the third year it was evident to me that I probably had not made a wise choice of majors. I began to accept the fact that the cattle industry was one in which having a network of friends and working associates was a must. One older professor stood before our class and announced that one was either born into ranching or married into it: there was no other way to enter the industry successfully. I began to realize that my background was such that business ownership or ranch ownership would not be an option for me.

At the same time I began to understand that there were many careers that were just as interesting and far more useable and consistent with my heritage. It was a deep down uneasy feeling. I had money for four years - insufficient to change majors in mid-stream. I have since tried many times by various ways to get into production agriculture but just could not fulfill that dream. I can only assume on faith that such has been the will of the Lord for my life. Though I believe the Lord directs us in the way we should go, I also believe that along the way He allows us latitude of decision and choice. I believe He expects us to make wise choices according to our spiritual maturity.

I knew that upon graduation, my educational deferment would disappear and there was the distinct probability that I would have to enter the Korean Conflict as a member of the United States Armed Forces. Regardless of the looming draft, I had a growing desire to go to Latin America as an agriculturist.

What was God's plan for my life? Growing up there was seldom any discussion regarding the positive influence of Christianity and my belief in God upon my life. Discussions usually revolved around the negative perception of what my Christianity would allow me. I think the proper perspective is to ask, how do I live out my Christianity?

Later in life I realized I was carrying around a negative attitude reflected in the phrase, 'I can't do that because I am a Christian.' Since that epiphany, I have begun to reflect, to try to live with a more positive approach that says, 'whatever you do, do all to the glory of God.' For me the difference between the two attitudes is a deep dark chasm.

Along with my studies, I learned a craft in my years at Cal Poly that brought a great deal of satisfaction. Another student, who was a friend, urged me to enroll in a leather tooling class in Adult Education. I learned quickly and began to tool belts and other things for students in school. The craft turned out to be quite a moneymaker for me. Unfortunately, it took far too much time from studying. The struggle between the two was sometimes great because I certainly enjoyed both.

Dorm life at Cal Poly in the late 1940s and early 1950s was different than that found in universities today. There were no co-ed dorms as women were not allowed to attend the school at that time. I remember one humorous incident that does not speak well for my level of maturity then.

At Cal Poly during the time of my attendance, there were several types of dormitories: my first year was spent in rented barracks. The next three years were spent in dorms that were built during the war to house U S Navy student pilots while they trained there. The buildings were nothing more than concrete slab floors with walls of three-eighth inch plywood nailed to two-by-four framing. I think there were some sixteen rooms in each building with a hallway down the middle, a door at each end of the building and all doors in the building opened inward. Bathroom facilities were placed in and at the center of each building.

There was a tendency for students to choose dormitories in which others of their academic major lived: one or more dormitories was taken up by engineering students, another dorm was taken up by rodeo men... the cowboys, etc., but my dorm was a mix of several majors. The accepted *rodeo dorm* housed a rather wild group of men. One evening when all were assumed to be quiet in their rooms studying, several men felt the need of excitement: their chosen target was one, Cotton Rosser. They very quietly filled a thirty-gallon trash can with water, placed the can leaning against Cotton's door. One of the group went to the dorm next door to place a call to Cotton Rosser in his dorm. One person of the group answered the phone and then called Cotton to the phone. As Cotton opened his door thirty gallons of water poured into his room. Mission accomplished. They all dragged out the mops and buckets to help clean up.

Toward the end of the year, there were always parties and wild doings. One night, two men from my dorm came in quite late making a lot of noise, waking most everyone. I had to get up at 4:30 a.m. to go to work and the noise did not set well with me. I figured it was pay-back time. OK, I have since learned that pay-back is not a very enlightened nor profitable concept.

At 4:30 that morning, I lit the fuse of one very loud firecracker and threw it under the door of the offending residents. The noise was indeed ear splitting to all but the residents of that particular room. They slept on. I had one more firecracker. I lit the fuse and threw it under the door—again a very loud explosion without the desired result. Rather disappointing. Late in the day, a resident at the other end of the hall asked if I knew who had lit the firecrackers. I, of course, knew nothing about it. I feel sure they reasoned correctly that I was the only one getting up at that hour.

I maintained a Christian witness by virtue of the encouragement of Christian friends, some were my own age, some older and WW II veterans. I, with a number of other men and the help of a local pastor, started the first Inter-Varsity Christian Fellowship group on campus. On two occasions I attended IVCF retreats held in the Santa Cruz Mountains. I met men and women from Stanford, Berkeley, San Jose State, San Francisco State and other prestigious schools. It is difficult to convey what I felt at meeting these men and women as I witnessed their level of spiritual maturity. I learned there was a whole new world out there well beyond my imagination. I have often wondered what I would have become had I had their influence on my life.

Our IVCF group was a registered group on campus with all the privileges and responsibilities of such an association. At that time we were the only on-campus Christian group ministry at Cal Poly. Today I am sure there are at least several more.

When the yearly campus-wide Poly Royal celebration came around each April, our Inter-Varsity chapter was given opportunity to participate in the all-campus Carnival evening. We were a small group and not capable of sponsoring the large shows, exhibits or games. So we chose to take responsibility for the sale of tickets to get in and to play the various games and see the shows. After all we were Christians belonging to a Christian club, wouldn't that guarantee total honesty while handling the money taken in? Wouldn't the other clubs trust honest Christians to handle their money for them? Fortunately, that is the way it worked out. As secretary/treasurer for the IVCF, I was elected to count and transport the money taken in at the end of the evening. After the Carnival closed for the evening, we counted $1200 which we left in a secure office for safe deposit. We had a lot of fun during the three years we participated.

Though I have used the vocational education I gained very little, I feel it was all worth the effort and money spent. I did use much of the vocational education the six and a half years we were in Ecuador. It was a joy and I thank the Lord for His kindness in allowing me to do so. The college education experience opened whole new worlds to me, opening and expanding vast areas in which my curiosity could explore to its heart's content.

As graduation drew near, I began to think more seriously about my desire to get married. I wanted to share my life with a wife, someone who was a bit different. I was looking for a girl who could live a solitary life and yet adapt to unusual circumstances. Most of the girls I knew in Santa Barbara were prepared to settle down near family. They did not any education beyond high school. Marrying a woman with a college degree was important to me.

One evening, I attended an event that included a brass quartet from Westmont College. A young woman playing the trumpet caught my eye. She was a bit shorter than me with curly black hair and a confident smile. After their part in the program, she stuck around rather than walking out the door.

Later on, I was to attend an event at Westmont College and in conversation with other young men, the fact that I had attended Cal Poly rather than a Christian college was revealed. Two men seemed appalled by the idea that as a Christian, I had not attended a Christian school. This attitude was not uncommon among the Christians of that day. The young lady with the curly black hair did not demonstrate that attitude!

As Valentine's Day drew near, I had two tickets for a banquet in Santa Barbara and no date. Through a friend, I asked the girl with the curly black hair if she would go with me. There was no time to receive a reply. Unlike today, long distance calls were for emergencies.

That evening, I borrowed Derral's 1934 Ford, a medium blue car that could really move. I drove down to the boarding house where Maxine Vaughan lived. She apparently found my invitation acceptable and we enjoyed our evening together.

Sixty years later, at another Valentine's Day banquet, I pulled out the two tickets for that first evening. She never knew I was such a romantic nor that I had carried the tickets for that first dinner all those years in my wallet.

Our courtship was carried on from a distance by way of letters and weekend dates. Westmont College and my home were both in Santa Barbara, so I went home almost every weekend to spend time with Maxine. She was too far away to just go and talk things over, to get to know her through frequent conversation. At the time we met, both of us had recently come out of break-ups with previous boyfriend- and girlfriend-relationships of long standing and it would have been much better had we been able to spend more time with each other.

Maxine graduated from Westmont in June of 1950. With a teaching certificate, she was hired to teach at Earlimart Elementary School in the mid-San Joaquin

Valley. During the school year of 1950-51, our courtship was made more difficult due to the long trip from San Luis Obispo to Earlimart.

As Maxine and I dated and became better acquainted with each other, we each had ideas about what we wished to do in the future. Early on Maxine had committed her life to the Lord for service in missions beyond the borders of the United States. China was a strong possibility. Due to the political situation in China in the 1950s, entrance into that country as a missionary was out of the question.

I had a strong desire to work in Latin America in agriculture and at the same time, work in the local church. No other part of the world was of any interest to me for such work and life. As we considered marriage and a life long commitment, I made it clear that this would be a partnership with Maxine helping to make all our decisions. After much thought and prayer, Maxine began to consider working in Latin America. She understood that I was strongly com- mitted to agriculture and a rural life.

As we talked about marriage and what a life together would look like, we obviously had to factor in the current war in Korea. It was a foregone conclusion that I would have to serve time in the military before doing anything else.

Eventually our engagement was announced which was a surprise to both our parents as over the previous two years we had broken off our courtship once or twice. Both sets of parents seemed happy with the announcement. My mother seemed especially happy because she thought Maxine would be *good for me*.

Maxine and I were married the 9th of June 1951

in Draper Chapel of the First Baptist Church of San Diego. Our honeymoon began to reveal the first evidence of our cultural differences. We honeymooned in the campground of Huntington Lake, east of Fresno, California. Maxine had lots of experience camping; her parents and family were frequent campers. Camping was not the problem, camping on a honeymoon was. In the early 1950s, a camping honeymoon was not common. We did seem to have a good time. The following summer we returned to Huntington Lake for a week of camping.

When I finished school in August of 1951, the Korean war was not over and the draft was still pulling men into military service. I received notification that I was now classified 1-A: eligible to be called in a draft at any time.

One evening in mid-September, Dad called to tell me that I had received notice to report for the draft in October. He asked me to join the Navy rather than go to the Army via the draft. Though the period of service would be almost twice as long, I thought it would be worth it. I quickly returned to Santa Barbara to enlist before the draft board could pull me into the Army.

In mid-October 1951, after only four months of marriage I was inducted into the U S Navy. I left Maxine in Earlimart so she could continue teaching, completing her contract. We didn't know it then but such leaving and returning was to be the custom of our lives together for the next four and a half years. Our courtship time had been carried on through letters and weekend dates. Now living together during marriage was sporadic based on my assignment, certainly not sufficient to get to know each other.

College Bound

The summer of 1952, Maxine had completed her two-year contract in Earlimart. By then assigned to the Hospital Corps School in San Diego and we knew that home-life while in the Navy was uncertain at best, I was pretty sure that San Diego would be home base. Maxine's parents lived there as well, so it seemed best that she live in San Diego. Teaching jobs were easy to find in San Diego and she settled into one teaching third grade.

Maxine looked much younger than her 22 years. One evening after school she came home upset but still able to chuckle about an incident with the mother of one of her students. The mother had come in after school for a conference with Maxine concerning her child. She had never seen Maxine and didn't know what she looked like. She entered the classroom, looked around and asked Maxine where she could find the teacher. Maxine informed her that she was the teacher. The woman was so amazed at 'the age of teachers nowadays.'

As I remember, that period was not entirely a happy time for either of us. We did have some very good times but the adjustments and stress were difficult. Teaching was not an easy job for her. After I returned from the east coast with a rank of Ensign and was assigned to shipboard, I was freqently arriving at or leaving port to go out to sea. We never knew when I would be in and back out. I suppose that is the life of most military personnel, especially those in the U S Navy. The military is just a difficult place for a young married couple to begin to build a marriage. The fact of such difficulties was born out a couple of years later when our ship spent the summer in Puget Sound of Washington State as there were thirteen divorces among the young married shipboard personnel over that three month period.

We estimate that over the period of our first five years of marriage, we spent only about two and a half years of it together. No wonder it was difficult to get things right the first few years after I was discharged from the Navy.

I wonder what it is about serving time in the military that has such an effect on a person's life. Not everyone is affected in the same way from that experience. Maybe military life reinforces one's basic personality. If one is naturally and highly disciplined, then certainly the military training enhances one's disciplined nature. On the contrary, if one abhors a disciplined life one will abhor it even more yet after spending any time in the military. I think both the concentrated and disciplined training combined with the shock of moving from what we know to an unfamiliar community has something to do with the permanent imprint on us. The process is almost violent. There is little time to stop and consider the options that confront us. The training becomes a matter of survival until one learns to navigate the military way of living. The military does get one's attention.

I enjoyed my time in the United States Navy. I received an education in many areas that I would not have gotten anywhere else in the world. The four and a half years I spent in the Navy certainly has influenced and will continue to influence my life—for good and for ill.

I do not remember much about getting on the train here in Santa Barbara, destined for San Diego. Boot camp was like nothing I had ever experienced before, nor was I adequately mentally and emotionally prepared for it. In keeping an open mind, I was still appalled at the low level of social skills, the lack of ability to read and write even on an elementary grade level of the other recruits. Midway through boot camp I was writing letters for those who couldn't write: to girl friends, to parents and other letters of general need. I was helping men read, helping them learn to read. I was amazed though I shouldn't have been, I had listened to men from WW II talk of boot camp. As I remember the matter, I believe most of the men in my company were from the southern states.

There were 90 men in my Company. Our Company Officer was a man with a Chief Petty Officer rating. He complained he was tired of the Navy, tired of military service and wished dearly that he could just go home and stay retired. After twenty-some years in the service, he had been called from Reserve status to serve a few more years during the Korean Conflict.

Out of 90 men, at 22 years of age, I was one of the three oldest men in the company. The men in the remainder of the group were between 17 and 19 years old. There were several with IQs under 98. I was the only college graduate in the bunch.

While in boot camp all college graduates were encouraged to apply for Officer Candidate School (OCS). Candidates got one chance to make it through—no second chance. Upon graduation, candidates emerged with a rank of Ensign. The application included the completion of many forms and questionnaires followed by a 40 minute interview before a Board of three Naval Commanders. I was so stressed out I don't know how I kept from breaking. Uncertain of how to act before a Board of three Commanders, I decided the best stance was one of absolute respect and honesty with a good bit of reserve. I made it through and then wondered how I did it.

The wheels that process such applications grind exceedingly slow. Not only had I applied for OCS as an alternative, I had chosen to pursue a Hospital Corpsman rating. Hospital Corps School was 16 weeks in a training center. I was assigned to the Great Lakes Naval Station Hospital.

As I waited for the assignment, the men and women I worked with were several rate levels above me. Only one thing seemed to occupy their minds: parties with sex and liquor. All were either single having never married or married and then divorced. I expressed the fact that I was happily involved in a young marriage and intended to stay that way. I expressed my disappointment at being assigned to the Great Lakes Naval Training Base. They asked me if I indeed did *not* want to go to Great Lakes and I confirmed that fact very clearly. Neither they nor I said any more.

A few days later as I packed, I was suddenly pulled aside and told I could *not* go. My pay records were missing from among my service records—and without my pay records I wasn't going anywhere. I realized then why the ladies at the financial office had asked *very clearly* if I really did *not* want to go to Great Lakes. They had slipped my records out and hid them so that I could not be transferred. A day or so later my pay records showed up—of course, it was too late for Great Lakes. I was transferred to Corpsman training at San Diego Naval Hospital Corps School. I often wondered if they thought that was the least they could do for someone who was trying to be a good husband and a good officer. In retrospect, I realize that God had a purpose for my training as a Corpsman, experience I would use years later in a remote area of Latin America.

In December of 1952, I finally received orders to OCS in Newport, Rhode Island, reporting for school there in January 1953. Our class was made up of 1,300 men. Only 833 completed the course. It seemed to me that the whole course was a contest between the administration and the officer candidates. They tried to make life so rough one would quit, fail academically or break enough rules to get thrown out. Twice I was placed on the weekly academic probation list. Many men just quit, feeling it below their dignity to put up with the daily little hassles and harassment. Many could not make the academic grade and of course there were a few who could not live with the rules and regulations. A number of men caught colds or the flu, which made study very difficult for them so they had to drop out.

There was not much time for hands-on training in OCS. We got some hands-on time in navigation classes using sextants, various compasses and hand instruments used to lay out courses on charts. Within several years the only instrument in use from those days was the gyro compass. Everything else was traded for technologically advanced instrumentation. There was also some hands-on training in shipboard damage control. That involved the shoring up of hatches opening to flooded compartments...the flooded compartments were real and were wet!

There had to be some classroom training in gunnery and engineering practices. But the 1950s was a period of transition as the military was moving from mechanical guidance devices to electronic devices and so much of what we learned was obsolete by the time we got on board ship.

After graduation I was ordered to the Stockton Navy base. While we were there our daughter, Ruth, was born at the Oakland Naval Hospital. I moved my family to San Diego to live near Maxine's parents after I was to report to an LST—Landing Ship, Tank 602— to replace the Engineering Officer. I already had been introduced to the Amphibious Forces of the Pacific Command while serving on the PCEC. Now reality hit home. We were badly under-staffed and therefore over-worked. The engineering department crew should have numbered 25 men. In my three years aboard ship, we never had more than 15 men. Due to the workload, morale was always a battle.

As an officer I was expected to sink or swim in leading a group of men doing a job without first knowing either what they were to do or how they were to do it. The responsibilities assigned me by the captain included maintenance, preventive maintenance and running all the machinery on the ship. The PCEC had two large diesel engines for main power and another that drove the electric generator. The LST had two main engines, three generator engines, two engines driving pumps and a large number of electric motors and systems. I had the responsibility for four landing boats as well. All I knew about diesel engines was that they are internal combustion engines that did not use spark plugs to ignite the fuel as gasoline engines do.

I had 12 or 13 men under my supervision and at least 10 of them knew more than I did about the machinery on the ship. I became a learner and at the same time their boss. I used this Navy *leadership training* to good advantage in later years.

Chief Petty Officer Engineman Vargo was a tall man, physically powerful and a loner. He was a highly disciplined man in all areas of his life except for the times he went ashore — then he turned into a common drunk. He was obviously an alcoholic; he knew it as well as everyone else. When he first joined the Navy in the late 1930s, his first stint aboard ship was as a coal-shoveler on a coal-burning, four-stack destroyer which sailed up the Orinoco River. He had been through World War II. His LST had served in the Mediterranean theater making the Anzio Beachhead and then on various beaches in the Pacific Theater. At Anzio his ship sustained considerable damage in the landing. But they got their cargo of battle tanks and crews off the ship and the ship back off the beach to make other landings later in the war. He knew diesel engines like no man I have ever known.

He and I had some long talks about his behavior and about the Lord. He knew all that I told him. Not long before his discharge while we were in San Diego, he went ashore for the evening. The ship was to sail for San Francisco the next day. He had not returned to the ship by the time we sailed and so was listed as AWOL. Some 72 hours later he showed up in San Francisco. The captain of the ship told me he would not court-martial the man as he had put in almost 20 years of hard service. He did not want to mar the man's record over such an incident. Such punishment would hardly be justice after all he had gone through. The captain did however, disallow him going ashore again until he was successfully released with an Honorable Discharge some weeks later. Chief Vargo stood before us at attention as the captain and I told him of our decision. There were tears in his eyes. In spite of his faults and weaknesses, I had a great deal of respect for his loyalty and his ability.

We made several trips from San Diego to Japan, most entailed rather boring days and nights. As engineering officer, I had oversight responsibility for the machinery on the ship as well as having to stand duty as officer of the deck with a schedule of four hours on and eight hours off.

The first trip we returned from Japan on the PCEC, stopping at Midway Island to take on fuel. Midway Island, one of the most beautiful of the Pacific Islands, was the site of one of the great and decisive sea battles of WW II. There is a coral reef that completely surrounds the island except for one narrow channel. Small ships can pass through the treacherous channel into the inner waters. We made the passage with a great deal of care and tied up at a small pier to stay for about five hours. With permission to go ashore, I walked to the far side of the island and went swimming. The water was warm and crystal clear. I could see the bottom as if it was mere inches deep. Relics of WWII remain visible here and there on the island and in the water.

As we backed the ship out of its position next to the pier, a loose piling damaged a screw. We made a repair stop in the shipyard at Honolulu, Hawaii. That week was a marvelous time of rest and recreation for the crew.

As we crossed the Pacific with another ship, the weather remained good. Stationing the ships about 50 yards apart, we put life boats in the water and allowed the two crews to go swimming. After the wonderful interlude, I learned that the ocean depth was around 25,000 feet deep at that point. A 10 pound ball falling at 10 feet per second would take 41 minutes to reach bottom at that point in the ocean.

On a second crossing three years later, while serving on the LST 602, our squadron of five ships endured a constant stream of oncoming winter storms. Unlike the previous crossing of 30 days, the storms added two days to our voyage.

At one point the opposing winds were blowing at some 80 knots. Our engines were running at full speed yet we only made about 1 to 2 knots headway against sea and winds. The swells came at us so high that when in a trough we could not see the other ships that were one quarter of a mile away. We estimated the swells to be around 60 to 80 feet in height. The voyage was a time of great intensity bringing on immense fatigue in our concern for the well being of the ship against such seas and weather. Every step I took was measured against the constant rolling and pitching of the ship. In order to rest and not roll out of our bunks, each man slept on his stomach with a knee placed between mattress and bunk-side. The crew was made up of many new recruits and some were seasick for the entire trip.

The mast on one of our sister ships snapped about 15 feet from the top, losing much of its electronic antennae and radar gear. It was a very scary incident.

In May of 1955 our ship received orders to proceed to a private shipyard in Seattle, Washington where we were to prepare the ship for a trip into the Arctic Ocean. The Distant Early Warning Line (DEW Line) was a series of radar stations above the Arctic Circle from which the U. S. military forces could detect any sort of airborne attack from the communist world. About forty ships took part in the project; both leased private ships as well as U. S. Navy units.

While in the shipyard, the Navy installed new experimental screws (propellers) made of a titanium metal on our ship, hoping they would not be easily damaged in the ice we would encounter. Heavy timbers were attached to the sides of the ship at the water line and the timbers were covered with metal sheets to protect the hull of the ship against puncture by ice. We took on additional refrigerated food storage units, barrels of fuel and a helicopter-landing pad. We loaded two earth movers, other vehicles and cargo into the tank deck and left Seattle in August with the rest of the convoy.

The project was to deliver supplies to the shores on the north side of Alaska and Canada. The north Pacific that year was not a pleasant place to be as we encountered heavy seas all the way to Point Herschel where we stopped, anchored and waited for the ice to move out from shore before we could proceed to Point Barrow. While steaming through one particularly rough storm, we picked up a message from another LST, 100 miles away. The storm had caused several cracks in the metal in their decks and they were afraid the ship would break apart. From then on I made frequent inspections looking for similar problems on our ship.

One afternoon while waiting at Point Herschel, I looked out to sea and there on the horizon I saw what looked like a solid wall of ice some fifty, maybe a hundred feet high. Quickly I realized it was a mirage of what was over the horizon, not a wall of ice, but an ice field. The mirage stayed some 45 to 75 minutes and then disappeared as the atmosphere changed with the onset of the evening. There was an eerie atmosphere about the whole scene.

As the ice cleared, we moved around Point Barrow and on east. Several times while the ship was beached and unloading cargo, I took walks up on the tundra or down the beaches. On the beach I found whale bones and strange rocks. On the tundra there were lichen-covered rocks, strange flowers and plants. I encountered several marmots so fat they waddled along.

At several landings, Alaskan natives were camped for the summer. They were curious about our operations. Our landing boats transported cargo from larger ships to the beach. Of the forty some ships involved in the operation, our ship was the first to round Point Barrow, indeed the first ship of its size to ever round the Point and sail east.

Our ship had been built to operate in shallow waters, for landing on beaches and backing off when unloaded so we were able to navigate in very shallow and dangerous waters. At times the Arctic waters were so shallow our propellers kicked up mud in the wake. At another time we were sailing at night at five knots when we hit a large ice floe that was grounded in shallow water. The collision smashed our bow doors which made them inoperable. Fortunately, we were on our way out and all cargo had been unloaded. There were two United States Coast Guard icebreakers along with the convoy. Their draft of some 30 feet kept them far out from shore and inaccessible to help us if and when we became stuck in the ice. In one case we did get caught in the ice. The ship and ice were blown by wind toward the shore. We spent some four hours attempting to extricate ourselves as the icebreaker could not help us. In another incident we were caught in ice in deep water. In the attempt to break us free, the ice breaker pushed a block of ice into our screw that bent the 5-inch diameter shaft, making it inoperable for the rest of the trip.

The Navy Underwater Demolition teams had surveyed the beaches where we were to beach and unload our cargo. Unfortunately they missed an underwater rock at one beach. As we were sailing west one bright day heading home, we heard something scrape the bottom of the ship. A few seconds later a couple of sailors fairly flew up the ladder out of the auxiliary boiler room. Water was coming in fast as we had run over an uncharted reef of sharp rocks, causing a split in the bottom of the ship. The boiler room filled with sea water to the water line. We lost one fresh water tank and one fuel tank to the inflow of salt water. We were able to put a patch on the hull of the ship over the split, pump out the water from the boiler room and weld the split shut. Such were the dangers involved.

I have been thankful to the Lord many times for my Navy career as it was a time of learning, a time to form good habits, to set priorities and goals. I learned much about myself, about my weaknesses as well as my strengths. In May of 1956 I was discharged from the Navy.

I have done a great deal of thinking about our early-married years. As I said previously, during the first five years of our marriage, we were together only about two and a half years of that time. We had some very good times. Our two children were born in U S Navy hospitals: Ruth at the Oakland facility and Garth in the

List of My Favorite Things

Favorite places to be on earth:
West side of the Sierra and the Rocky Mountains.
The range lands of Wyoming and Montana.
I do not like southern California.

Some of the odd places I've been:
Swam in the waters off Midway Island, it was beautiful.
Swam in the waters of the middle Pacific Ocean
where the depth was 25,000 feet.
Walked the deep eastern jungles of Ecuador.
Walked the shores of northern Alaska
and Canada on the Arctic Ocean.

Some of my favorite things to do:
Ride my horse in the solitude of the mountains.
Riding my motorcycle on most any quiet road
winding through the mountains, valleys and meadows
of the western United States.
Plant and tend plants of various sorts to watch them grow
and take shape as they will.
Talk with and question older people about their younger lives.
To watch weather storms as they pass overhead...their fierce
winds, pounding rain, thunder and lightning.
As well, I enjoy sitting on a grass-covered hillside on a quiet
afternoon, listening to a spring breeze wafting over a field
of wild oats or any tall grass, watching the seed-heads sway
in the breeze.
I enjoy the silence of the desert, the hills and the mountains.
Some think the sea or the beach is a place of silence, it is not.
The beach is noisy with breaking waves
and movement of the ocean...always the moving ocean.

San Diego hospital. The day Garth was born, I was far to the north in the Arctic Ocean over the top of Alaska. The afternoon of the 26th of August 1955, I was standing on the bridge of my ship watching our slow advance through the ice floes. A radioman called me to the radio shack to watch the notice of Garth's birth being typed out on the Teletype machine. Quite a thrill—almost like watching your child being born, though I am sure Maxine would not agree this was a true *likeness*.

Maxine bore the brunt of caring for the children in their first year or so of life. Six months or so after Garth was born, Maxine noticed his growth progress was not normal. The doctor she was using at the time, suggested Garth might have cerebral palsy. This was confirmed at nine months of age, shortly after I had been discharged from the Navy. I will never forget his advice to us at that time.

"You can search the world over for a cure, you can spend thousands of dollars and you will find, finally, there is no cure. All you can do is begin now with therapy to help the bones and muscles grow as correctly as possible. Treat him as normally as possible. Don't build family life around him and his disability."

Without regret, we have followed the doctor's advice. Little did we understand the impact his disability would make on the rest of our lives. We now look back and realize that attention to his needs have pretty well governed the parameters of our lives. Without doubt we see where the Lord led, intervened and directed our lives.

Over the years it has been interesting to watch Garth deal with his life as a mentally and physically handicapped person. I suppose that most handicapped people deal with life in much the same manner, especially if they have been so from birth. I don't remember how many but Garth had numerous operations to correct physical deformities—operations on his eyes, hand, foot and leg. Cerebral palsy has affected all four limbs—his right side is the most severely affected. He has over-come and progressed in growth and development over his handicaps by sheer force of will power. Over the years he has demonstrated this capability in many areas of his life.

My deep desire to go to Latin America kept me from making a career of the Navy though many commanding officers suggested I do so. Throughout those years, my wife Maxine was my companion and partner. She both suffered and enjoyed some of the things that I did. The one part she most disliked was knowing that I was seeing some of the world while she was staying home in San Diego. I must agree with her that it did not seem fair.

CHAPTER FOUR

No Coincidence

My desire to work in Latin America after my years in the US Navy led us to the middle of the desert, back to where I had taken my first breath. Anticipating my discharge, we researched graduate level programs with an emphasis on doing business in Latin America. As I did not see myself as a missionary, preaching the gospel, I rejected the idea of Bible school or seminary. I did not feel that I had received such a call.

We applied for and were accepted into the American Institute for Foreign Trade (AIFT), a post-graduate school north of Glendale, just outside Phoenix, Arizona. In late September, we moved into on-campus housing at AIFT. Our apartment was one large room that served as living-dining-family-bedroom. As far as we could learn, we were the only Christians on campus. Many of the men students were veterans of the Korean Conflict and had much in common. Most of the men and women on campus were acquainted with the world of commerce. Maxine and I came from a different world with different aspirations that set us apart just a bit.

The material we learned and the people we met, thrilled both Maxine and me. As we look back on the year we were at AIFT, we both must conclude that it was a difficult time but a good time. To some extent we felt *watched* by other married couples as we struggled with Garth and with the therapy he required for cerebral palsy. Garth had a smile for everyone, a wonderful attitude and made friends with anyone while in the constrictions of casts and braces.

In June of 1957, I graduated with a Bachelor of Arts in Foreign Trade. Maxine graduated with a Certificate of Class Completion. She had attended three key classes: both conversational Spanish and Spanish grammar, and a class in Latin American Area Studies.

We had hoped after graduation I would be able to get a job in Latin America as an agriculturalist and with this as a base we could work with a local church body as lay-missionaries. I found the agricultural production jobs with north American companies had qualification requirements that I could not hope to fill and none were willing to train an entry level employee. I kept in contact with several classmates from AIFT with the hope that maybe one of them might recommend something to us. Finally a letter arrived from a fellow in Panama with the recommendation that I contact a certain businessman there.

After some correspondence with the business owner, we agreed that I visit Panama at my own expense for an interview and to see their operation. The trip cost $259. which was all of our savings.

I left Phoenix by bus and traveled to Nogales, Arizona. Then I walked across the border into Nogales, Mexico to take a plane from there to Panama. The airport was a small strip with no terminal or control tower. Standing there, I watched the Douglas DC3 plane carrying an airline insignia land and take off, repeating the procedure several times. I asked a fellow who appeared to be an agent what was going on. He replied the pilot was practicing. Well, okay, practice is good. But when the plane landed and taxied to the loading ramp - *my* loading ramp, I had other thoughts! At such times, one is given to much prayer beseeching the Lord's intervention in any untoward event!

After a night in Mexico City, I caught the next flight to Panama. Though it was a larger plane, like the day before we seemed to stop in every country between Mexico City and Panama City except Belize. I was amazed to see the difference in country-mood that each national airport demonstrated.

Guatemala City's airport was heavily guarded with police carrying machine guns at the ready. Costa Rica, on the contrary was laid back and comfortable. When we landed in Panama City it was about 10:30 in the evening and I had no idea where to get a room for the night. Sharing a cab with another AIFT graduate, we checked in at the Panama Hilton which charged $20 or broken down, $3.50 an hour to sleep. This was way beyond my price range. The next morning the family I was to visit helped me find more affordable accommodations. My first night there, an intruder sneaked into my room through the unlocked balcony doors while I was in the bathroom. When I stepped out of the bathroom, he immediately ran leaving the money he was looking for. After locking the balcony doors, I went to bed not sleeping much the rest of the night.

The following day my contact took me out into the Panamanian countryside to their home where we talked and saw the countryside. We both came to the conclusion that I could not do an adequate job for them. I was not the person to manage their agricultural store in

Panama City. The next day they took me to the airport and I returned to Nogales, Mexico.

Back in Nogales, I hitchhiked a ride to Phoenix with a fellow traveler where Maxine picked me up. I was glad to be back home, but having spent our savings with nothing to show for the experience left me with a terrible empty feeling for some days afterward.

I believe that God wants us to dream and set a vision for our lives, to set goals for the future. We are to do the very best with the mental and physical abilities He has given us. Yet for some reason, He did not see fit to allow us to continue in the direction we had chosen. He had other plans for us.

With our savings gone, I went to work for a cousin in Glendale. He needed some drafting done for new cotton gins he was building for his employer. The company owned 11 gins over a 75-mile distance from Casa Grande to the Harquahala Valley so we used a small airplane to fly from one gin to another. They had both short-staple and long-staple gins. My cousin started to teach me to fly during that time. I made it through ground school and had logged 11 hours of solo flight when the company owners learned of our efforts. They ordered us to stop using their plane, explaining that insurance rates were too high. I so enjoyed flying alone on early summer Sunday mornings, the flights out over the desert then back home in time for Sunday School and church.

My job was to help re-design the long-staple cotton gin. We completed the project and sold six or eight gins to a cotton ginning company just south of Las Cruces, New Mexico. I was assigned the job of hauling the unassembled machinery to the site and building and installing the gin machinery. We loaded an old International truck with all the tools, parts and equipment. When I arrived, the gin manager immediately called me into his office. He sat down at his checkbook, explaining that though I was being paid by Community Gin Company, he was going to give me a check for $100. I protested that I was well paid. He said that my wages would not pay for the extra wear and tear on clothes, nor the emotions of being away from my family and handed me the $100. At the time Levi jeans cost $2.50 a pair and work boots were $15. He was one of the most generous fair-minded men I had ever met. I finished that job in about three weeks and returned home.

In the latter 1950s, the machinery in a cotton gin was relatively dangerous to work around. There were open and unprotected high-speed belts, augers and saws all without protective shields. There was the constant danger of fire. Carelessness was not something a gin worker could afford.

One day while working on the electrical system I ignorantly put my hand where I could not see it and touched a 440-volt wire. The jolt knocked me six feet back. In another incident I stood on a stool in front of a newly installed 220-volt electrical switch and breaker-box. I called for the main power switch to be activated and at that instant the box in front of me literally blew apart, blinding me for several minutes and knocking me to the floor. Careless workmanship had incorrectly connected the wires for a dead short.

I was working in the gin one morning when a fire broke out in a 14-inch auger ten feet above floor level. I climbed up on top with a hose, opened the auger trough and straddled the trough to walk along its edges and spray water on the cotton as it ran out to the end. My foot slipped and I fell spread-eagle over the top of the auger to keep from falling in. If my foot had slipped into the auger trough, I would have been killed. I am not sure that anyone could have done something more stupid than what I had done. Surely the Lord kept me from falling.

The men I worked with were men from the south who had come out west to find better jobs. Most were relatively uneducated, only a few had finished the sixth grade. Sitting with them at lunch or the few moments before work began, they told stories of events that could not possibly have happened but they swore to their veracity. Their superstitious stories were always told with a straight face. Then another man would break in with a story that topped the one just told. They were men a good deal older than I and they told the stories as much for my education as for fun.

In 1959, the cotton industry took an economic downturn and I realized my job would soon be phased out. Unknown to me, the owner of the local blacksmith and welding shop was watching my job performance. When I left the cotton gin, the man offered me a job. Eventually I became the shop foreman, a big job for a kid of 28 years. While interesting, the job paid only $350 a month for a 45 to 50 hour week without overtime pay. I learned much in the welding shop that I would use later in Latin America. I also learned more about people who were skilled craftsmen and uneducated but kind to me and each other.

There were several interesting men who worked there but ol' Slim is the one stands out in my mind. He was a superb blacksmith, tall and physically powerful, a smart man who gave his employer all he was paid to give and then some. He was an elder in the Church of Christ in Glendale and he lived by the principles of scripture as he interpreted them. Farmers came from far and wide to have him sharpen their plows, hard-face cultivator shovels as Slim was the best around. When needing a big or heavy piece of metal bent, they came to Slim. He worked inside the large front doors of the shop. People walking by often stopped to watch as Slim wrestled heavy pieces of red-hot iron on the trip-hammer.

And then there was Frank. He was a welder who told the whole shop that the day I came to work there he would walk out. When I showed up the first day, he did not walk out. He was a good welder but a 'know-it-all' talker. One day a dairyman of loud voice, strong words and power of presence called asking if Frank could be sent to his dairy for several days work on site. After Frank went to the job site, word got back to the shop that when Frank stepped out of the truck, the dairyman set the ground rules: "Frank, I will do the bossin', I will do the talkin' and I will do the cussin'… you will shut-up and do the workin'!" With those ground rules understood, the dairyman said Frank did a fine job. The other men at the shop chuckled for several days over the story.

As the hot months arrived each year, several of the Japanese farmers often dropped by with a crate of melons that had cooled over-night. They were delicious and one of the highlights of the job.

This was a time of dynamic agricultural, industrial and residential growth with all three competing for the same land and money. With such dynamic growth and competition, the shop was asked to build new and innovative machinery. On several occasions the shop built and installed livestock feed processing equipment at large and small feed-yards. We built and installed some mining equipment in the mountains east of San Manuel, Arizona. It was all interesting but hard work with long hours.

Across the street from the shop was a large livestock feed processing plant. Our shop did most of the welding and repair work for them. A young fellow about 18 or 20 years old, who was the son of one of our welders, worked for the feed mill as a truck driver delivering truckloads of feed to dairies and cattle feeding facilities. The trucks were equipped with a six-inch grate-covered auger in the bottom of the truck-bed. When activated, powered by the truck engine, the auger carried the feed to the front of the truck-bed to another auger inside a tube and dumped it into a designated grain or feed bin.

One afternoon, my boss hurried over, pushing me out of the hearing of our welders. He said, "Don, Manuel's boy was in an accident. Take our big service truck and run out to this farm. You'll need the cutting torch. Take your goggles and tools."

I hadn't the least idea of what I was going to find. As I got to the farm the ambulance and doctor arrived. I knew then it was serious. I drove up fairly close to the feed-truck and stopped. The farm manager showed me what the problem was. When I saw the boy, I had no time to get sick.

The young fellow had not totally removed the canvas covering the grain. When he activated the auger, the grain was augered out and eventually the canvas got wrapped up in the auger. The fellow could not see over the side of the truck what was happening so jumped up in the truck on the canvas, removed the protective grate and then his foot slipped into the turning auger. He was caught with the engine and auger running and no way to stop it, trapping him irretrievably. He had been told never to get in the bed of the truck while the auger was running and never to remove the grating cover from the auger while the truck was running.

The young man screamed and finally a farm worker heard him. Running over, he arrived as the truck engine ground to a stop. The canvas had so en-wrapped itself in the auger that the auger could no longer turn. The auger had drawn the fellow's leg in and under a piece of metal across the top of the auger. The metal had reached the boy's groin. Had the auger continued, it would have torn the leg from the hip socket. The boy was bleeding profusely, shock and trauma were taking a terrible toll. I was told to cut the metal from around the leg so as to extricate it.

The father arrived as I began to work. He could do nothing but watch and console his son. I was in and out of the truck-bed, cutting metal here and then back under the truck to cut it somewhere else. The young man screamed with pain both from the wounds and the heat of the cutting torch. The doctor kept exhorting me, "Hurry, Don, we're about to lose him."

His lower leg bone was cut just above the ankle and the doctor finally told me to cut the tendons that held the foot to the rest of the leg. I took out my pocket-knife and cut it. The foot dropped to the ground under the truck.

The ambulance crew, the doctor and a farm helper

lifted the boy out of the truck and onto a stretcher, carrying him to the ambulance. The boy lived but then died a few years later in jail. It was a terrible job. The doctor later commended me for how cool I had acted under such pressure and terrible circumstances. Neither of us would ever forget that day.

While I worked, Maxine dealt with doctors and therapists, taking Garth to and from appointments. Distances are great between places in the Phoenix area. Summers were so hot. They just get hotter sitting and waiting for traffic lights to change. Air-conditioned cars were available to those who could pay for them. Air conditioning was not in our budget. There were no freeways that were of any help to traffic.

Doctors were learning and trying all kinds of contraptions for treatment of cerebral palsy. Special shoes, glasses, braces, casts of all kinds and operations—Garth had them all. He endured with such a very good spirit and attitude. For him they seemed to be just another part of life to be lived through. Because of Garth's good attitude, doctors and therapists always enjoyed working with him, any attention given him was returned in full by way of attitude and cooperation. It seemed that everyone was his friend. The bills for Garth's surgeries and therapy were paid through several foundations and doctor's donation of his services.

Though we were working to pay our living expenses, Maxine and I still had our hearts and minds set on Latin America. From a spiritual viewpoint, the four years we lived in Glendale were filled with intense desire for spiritual growth—years of frustration because it seemed nothing was happening. I desired so deeply to go to Latin America in agricultural production. Every sort of opportunity that came my way that didn't point in that direction I rejected. I think that sometimes we want something so much that we are blinded to opportunities and possibilities that come along tangential to our main goals and priorities. Yet one ingredient of any sort of success is perseverance toward a goal, and I certainly did persevere. I remain convinced that God was with us in those years.

List of My Favorite Books of the Bible

Someone has said that the Bible is the Word of God couched in the language of man. I think that is true. I enjoy noting how God used words to communicate His Word to us.
The following list contains the books of the Bible that I find myself most intrigued with, have studied the most and return to time after time.

Joshua
The Psalms
Isaiah
Jeremiah
Jonah
Habakkuk
The Gospel of John and his three Epistles
Ephesians
Philippians

When we moved to Arizona in 1956, we visited several churches and felt comfortable in the First Baptist Church of Glendale. Reverend C. I. Tucker was the pastor. Maxine and I grew spiritually while there. Over the years, we had the feeling that we did not belong to the inner core of the church. Many people helped us deal with Garth's problems and they too grew, because they had never had to deal with people who were part of their church family who had a handicapped child as we did. As Maxine taught the children's groups and I taught the college and young married groups, we did gain a place in the church family.

Pastor Tucker seemed happy to have us as part of the church but I struggled a bit with his authoritarian style of leadership. At the time I was teaching a group of college and business young people. I struggled with the endeavor. One Sunday after a particularly difficult class I went to the pastor and told him I was going to quit. As I look back on that incident, his response was comical.

"Don, as pastor of this church, the Lord has not told me that you should quit and so I don't really see how I can let you quit."

I stuttered and stammered along for a minute or two, we talked about the matter and I went back to teaching the class. The incident was one of the best things that ever happened to me.

One day Maxine was asked to pick up the speaker for the Women's Missionary Society meeting at the church. They had a long drive over two lane roads from the woman's home to the church with many stop signs along the way. Moneen asked Maxine about our life's vision. Maxine told her of our desires to work in agriculture in Latin America.

In her response, Moneen spoke of her brother-in-law, Frank Drown, working in Ecuador. He was looking for an agriculturalist to help him in the eastern jungles there. She suggested we write to him and see what he had to suggest. The correspondence went well and we applied to the Gospel Missionary Union (GMU) under which the Drowns worked. They accepted us for an upcoming candidate period.

Throughout the life of Glendale First Baptist, a number of young men had grown up, gone away to Bible schools and seminaries and had become good and successful pastors. Young women had grown up and married pastors. But over the fifty years the church had been organized it had never had a young person from its membership enter the ranks of foreign missionaries. Pastor Tucker had never pastored a church member into that calling. We thought it kind of strange.

When Maxine and I announced that we felt called by our Lord to the foreign mission field in 1960, it was a shock to the church and it's leaders. We were the first missionaries to go out from the church but they just did not know what they should do about financial support for us.

Our stepping out in faith in spite of several inhibiting factors seemed to motivate and encourage other young people in the church to accept an apparent call to the mission field. It was quite exciting to watch. That is not to say the church hadn't supported missionaries before, they had. But none of their own number had become missionaries—we were the first.

The attitudes of the church leaders for the first six months seemed to be one of wait and see if we would actually made it to the mission field. If we did then they would support us. They didn't seem to understand that our arrival on the field would depend a great deal on their initial support. Pastor Tucker did support our going and our proposed work: agricultural development on the field. He told me that such an approach to missionary work should have been done long ago. He was a very practical man.

I have read and reread letters to and from our parents documenting and discussing what we perceived as our call to the mission field. It was a difficult time for our parents and us. Our parents seemed to think of us as misguided children, destroying their dream for our success. They didn't want to accept our change of their plans. The Vaughans had friends whose children had married and gone on to make good lives for themselves. Why couldn't we do the same. My parents were concerned for Garth's welfare and the opportunities for him in education and healthcare. They had hoped that I would use my education to make a good life for us. Maxine and I were getting involved in a life totally outside of their experiences.

In their lack of support, they did not believe our decision came of mature thinking. What support we got from friends in Arizona was a bit strained and superficial because they really did not know us that well. We had not lived in the community all our lives. We were both college trained and that also was something new for a farming community. Young people going out as missionaries was not part of their church experience. The months before we left for training were some of the most difficult of our lives together.

The time came for us to leave Arizona and go to

Kansas City for out candidate period with the Gospel Missionary Union. The period would last four months - September through December. We had disposed of most of our household goods, leaving what we would keep with friends. It was possible that we would find that we could not work with the mission—or that they would feel we would not be acceptable to them. The four-month period of time would give both them and us a look at each other.

We left Glendale in August 1960, arriving in Kansas City three and a half days later. My first impression of GMU headquarters was not good. By the next morning, we could have easily packed and gone back home.

I thought I knew all about heat having come from Phoenix: I was not prepared for the late summer humidity of Kansas City, Missouri. The GMU headquarters building was in a part of Kansas City that had fallen on ruinous days, headed for slum status. Women at the mission were not to go out on the street alone. Our children were not to walk home from school alone so Maxine had to walk to and from school with Ruth though the school was only a block away. We lived in two rooms on the second floor of dark, old building as part of the community, eating our meals in the dining room downstairs.

The confining nature and discipline of those four months put a real strain on us and our children. It was such a contrast to the life we had begun for ourselves in the west. We had just stepped into our thirties. We had spent four and a half years in the Navy, had made a home and life for ourselves in Arizona. We believed we knew how to handle problems of life at the mission as they presented themselves. The other candidates were young people just out of Bible school or seminary. None had any experience living on their own in the world. Ultimately, we decided we could put up with most anything for four months.

By mid-November, almost half of the original candidates had dropped out, had been asked to leave or left for medical reasons, unable to finish the period. Those remaining continued to wonder if they would complete their candidacy.

One morning I was approached by Mr. Don Shidler, the Director of the mission, and asked if I would be interested in driving him on a two-week speaking tour through Minnesota and South Dakota. He would pay all the expenses I would be the driver, using my car. Unfortunately, I was not comfortable leaving Maxine to face the staff on a daily basis regarding our children. Mrs. Shidler was especially hard on and critical of Maxine, as she was on most of the women candidates. My presence softened the edges generally. After talking about the opportunity, Maxine agreed that I should go.

The trip allowed me to learn to speak publicly and be introduced to people and churches who might be interested in supporting our work in Ecuador. Along with tutoring me in my knowledge of the Bible, through our conversations on that trip, I learned many things I never dreamed I needed to know. Mr. Shidler was a spiritually mature and wise man and I found our discussions always profitable.

I was absolutely unacquainted with the German Mennonite churches and culture, as well as other lesser-known groups of Anabaptist origins that we visited on the trip. Mr. Shidler and many members of the mission were Mennonites. All were good people and I enjoyed their fellowship. Our doctrinal differences existed only in minor and insignificant details.

We were visiting people in Iowa when Mr. Shidler directed me to a farm, the owner of which he said he needed to visit. We arrived to find both man and wife very happy to see us. After an hour visit I began to wonder about the husband, a rather gruff old farmer who apparently consumed maybe two packs of cigarettes a day. His wife seemed to me to be a very godly woman of great depth.

After we were out on the road, I asked Mr. Shidler how he dealt with people who smoked two packs of cigarettes a day and appeared to have gone as far with the Lord as they wished to go. How does he minister to such people? My question arose out of the context of my fundamentalist up-bringing, namely that smoking was a sin and indicated a poverty stricken spiritual life. I assumed that Mr. Shidler was of the same mind.

Mr. Shidler replied, "Don, I think the man is as deep spiritually as he will ever be. He is a fine man. He has made his peace with God. We should let him be, the Lord will deal with him."

His answer and his attitude answered a lot of questions that had been bothering me over the years. In that moment, I learned that Mr. Shidler seemed to be far less legalistic than the organization within which he worked. People and institutions change from within, not from without.

We traveled on to our destination, a church parsonage in Bridgewater, South Dakota. Mr. Shidler was to be the principal speaker at a missions conference there. I would give my testimony at a meeting. That conference was followed by a Thanksgiving service at another location, after which many of the congregation had dinner at a farmer's house.

The meal started with a huge bowl of chicken soup, plain, exquisitely tasting chicken soup. There was homemade bread and mounds of real butter on the table as well. I made the very sad mistake of believing that this was to be the sum total of the meal. What a mistake. After everyone finished their soup, the table was cleared. This was followed by more food than I have ever seen in a lifetime placed on a table to serve maybe 25 people. Along with a huge turkey, came platters of mashed potatoes, vegetables, several kinds of cranberry sauce, on and on. For desert there were pies and cakes and strong black coffee. I was so miserable by the time we got to our room, I commented to Mr. Shidler on my misery. He laughed and apologized for not warning me ahead of time to start out lightly on the soup and bread.

Mr. Shidler spoke the Sunday evening following that Thanksgiving Day. We left intending to arrive home for the annual meeting of the GMU Board of Directors. The men at the church knew of our plans and talked of a storm front that was moving in from the north, warning Mr. Shidler of the blizzard. After driving two hours, the snow began to fall quite heavily, building up on the road. Mr. Shidler saw a motel and told me to turn in. In my ignorance I suggested that I thought we could continue. We got the last room available. The next morning we found the lobby filled with people sleeping on the floor, in chairs or on couches. I remember a large semi-truck carrying a load of pigs parked across the street…they were not happy pigs! The blizzard was raging at full-force. Surely the Lord had taken care of us very well.

There was nothing to do but wait the storm out. Mr. Shidler suggested that we read through the book of Habakkuk. I thought, that won't take very long, only nine o'clock with the whole day ahead.

As we read, each taking a few verses, Mr. Shidler began to lay out an outline of the book, commenting on paragraphs while I took notes. We finished up about noon. I had never experienced such teaching in my life. That morning he opened up the book of Habakkuk to me and I have continued study of the book since that day. It is very difficult to express the profound impact he had on me that morning.

After lunch and a short nap, he suggested we read the Apostle Paul's letter to the Philippians in the same manner we did that morning. Again, he outlined the book for me, commenting on paragraphs. From that outline I studied and taught the book for many years.

That evening the storm began to pass and by the next morning the skies were clear. When we reached Sioux Falls, Mr. Shidler felt he should catch the next plane out for Kansas City and I would drive on alone, arriving late that night. When I think back on that incident, I feel sure that had I driven alone into the storm, I probably would not have come out of it alive. I did not have the experience to know when to stop and find shelter. There were some long distances between shelter and lots of places to slide off the roadway and into a ditch where one could freeze to death so easily.

When the time for interview by the board came, I believe we presented ourselves well. Mr. Shidler wanted both Maxine and I to be accepted. I have no idea how the board voted but I feel confident that the Lord worked through him. The Board of Directors of GMU accepted us into the ranks of their missionaries. They opened the door on a huge landscape of opportunity, of adventure, of hardship, of possible spiritual and personal growth. We could not imagine the expanse of the possibilities. When alone we just sort of gasped, "what have we done?"

With acceptance by GMU as missionaries we now faced the prospect of raising financial support from individuals and churches. The process of deputation and gaining financial support took us a few weeks less than a year. We didn't keep track of the mileage traveled. Maxine, Garth and I made one extended trip up through Oregon, Washington, Idaho and Montana meeting with individuals and churches. I made approximately the same trip one more time. I also made a trip back through some of the midwestern states.

As we struggled with public speaking and asking people we did not know for financial support, we struggled with the lack of understanding on the part of our families. We attempted to explain to our friends, Christian and non-Christian alike, what we were doing. My parents did much for us in caring for Ruth and carrying the load of board and room while we stayed with them.

After nearly a year we had enough support as well as money to buy clothing and other necessary household goods for the next five years. We began to buy what we needed, packing the supplies in crates. The crates were shipped about two months before our date of departure. We obtainedpassports, visas and other necessary papers and finally immunization shots. We studied flight schedules from Miami, planning to drive cross country to catch our flight south.

No Coincidence

One of the most difficult things for our parents in our leaving, was that we left just before Christmas, taking their grandchildren. Christmas Eve we arrived in Houston, Texas after driving all day. We peered through the darkness, looking for a motel. The four-lane roadway we were on was intercepted by a four-lane cross street and I was on the inside lane. Not knowing which way to go, I chose to go right across the outside lane which in general is an illegal turn.

After a half block, I saw the flashing red lights behind me. The policeman was understanding when I described our journey. I followed his directions to a suitable motel. After putting Ruth and Garth to bed, Maxine unpacked some Christmas decorations and made a Christmas tree out of the antenna *rabbit-ears* on top of the television set. She put out two or three presents each for the kids—the next morning we had Christmas before driving onward.

At the end of the five-day trip across the southern US, we arrived mid-afternoon in Miami. An older couple that made it their ministry to assist missionaries on their way to the mission-field. The man agreed to sell our car for us and send us the proceeds. Our flight was scheduled for about midnight. He took us to the airport before his bedtime and left us to wait with our baggage and tickets in hand. We had long since said our goodbyes to those who meant so much to us. Now there was no one to see us off. We felt as if we were four against the world, surrounded by people everywhere and not a friend in sight. It was such a lonely feeling and we took comfort in each other.

While Maxine, made waiting for the midnight flight as comfortable for Ruth and Garth as she could,

I paced the hallways, clock-watching, drinking coffee, wondering if we were at the right gate, wondering if anything would go wrong at the last minute. The flight was thirty minutes to an hour late. Foreign airlines had a reputation for working by their own clock. As I watched Maxine, aware of her patience with Garth and Ruth, she watched me aware of my impatience and anxiety about the next 12 hours of our lives.

Then the loudspeaker announced and a heightened sense of anticipation and tension gripped both of us. After all the work of the past year, we were passing the point of no return. Around us, the majority of the passengers were Ecuadorians returning home from their Christmas holidays in Miami. Most carried packages of all sizes, packing the items around them with no more room in the overhead bins. We had thought that the airline meant what they said when their instructions read to carry on board as little as possible. We were suddenly in the midst of Latin culture.

Arriving in Guayaquil, the GMU Field Director, Abe Dyck, met us at the airport. As he moved us through the process of customs, it was easy to see that this indeed was only the beginning. Abe was a very capable man in dealing with the customs people and used his connections for our time and financial benefit.

Guayaquil is a coastal port city, busy, dirty, hot and humid, situated up the Guayas River from the Pacific Ocean. It is a noisy city, especially at night. The city's outer edges are slum dwellings built on stilts over tidal flats. Little by little, the city pushes fill-material and dirt in under the dwellings and eventually they find themselves living on dry land. And the city grows larger and larger.

We lived with the Dycks for about two weeks while we completed paper work and waited for residence permits and papers. The Dycks, having never met us before, were very kind. For personal safety reasons, we could not just go out for a walk in the park to give each other a little privacy. Guayaquil is a city totally different from any we had been in, different rules, different ways, everything was interesting and some things dangerous.

In mid January, school started at the Alliance Academy in Quito. Early one morning, Ruth and I accompanied another missionary, who worked in Guayaquil, and his two boys to Quito. I settled Ruth in the GMU dorm, a boarding home for missionary children and enrolled her in school. Living in a home with 35 other children and two dorm parents was a new experience for her. I returned to Quayaquil with the promise that we would see her in two weeks.

Chapter Five

Macuma, Ecuador
with the Gospel Missionary Union

As we began our lives as missionaries in Ecuador I had two life-long values that formed the foundation for our efforts. I believe that God is sovereign and Jesus Christ is my Savior, Redeemer and Lord. For those who share that belief, stating such an elemental belief may seem unnecessary. I want to be very clear that this is where I stand. I also believe that God has formed me to a 'waterer, cultivator, or the expeditor' in accordance with I Corinthians 3:6. However far the cogitations and curiosity of my mind may wander, I always come back to those two beliefs.

From the time of our marriage, Maxine and I set goals and directions for our lives. Some people settle early into their groove of life and their groove becomes a rut that deepens and darkens as they finish their allotted years. Others never stay long enough in a groove to pick up any speed, moving from one shallow slot to another. Then there are those who broaden their initial groove to include all sorts of efforts and endeavors relevant to their interests. They look out upon an open vista of opportunity to learn and grow. Their lives become one grand symphony, rich and full. I do not like mediocrity, an attitude of 'that's good enough.' God has a plan for my life, a plan with latitude. I firmly believe that God has endowed me with abilities, strengths and weaknesses, all for the purpose of accomplishing what He planned for me to accomplish through those endowments.

The day came for us to leave Guayaquil and move to Shell Mera on the eastern side of the Andes, the place to be our home for the next year. Traveling from the coastal plain through the Andes Mountains to the eastern jungles is an awesome trip. The coastal plain in the early 1960s was covered with banana plantations. Two-lane highways run for miles and miles through them with only an occasional village or town to break the monotony of the scenery.

Large trucks carry the heads of bananas to a shipping point. The truckers line the inside of the truckbeds with banana leaves to prevent bruising and damaging the bananas. After each load is delivered they dump the leaves and rejected bananas at a selected site outside of the village or town. There the bananas and leaves rot while the local pigs dig through the mass, foraging for whatever they can consume. Passing by such a spot, one cannot avoid the 'vinegary' smell coming from such dumps. It is a smell one never forgets. I am reminded of it when I allow a banana in the house to ripen too much or open a bottle of banana vinegar.

Eventually the road begins the climb into the Andes on the narrow winding way with occasional vistas out over the jungle-covered mountains. The road climbs above tree line above where no trees grow except an occasional alien eucalyptus tree. Still climbing, the road enters the eternally cold and dry mountain paramo topping out at a pass at about 12,000 to 13,000 feet elevation. Mountains on either side tower 14,000 to 16,000 feet. Once over the pass, the road descends into the central valley corridor at about 8,000 to 10,000 feet, running north between the two rows of volcanoes on either side, some extinct, some active.

After a couple of nights at the GMU guest house in Quito we boarded the bus to Shell Mera. This little town on the edge of the eastern jungles reminds me of what I imagined to be a frontier town. Farmers bring the fruits of their labor to a public market, bartering with shoppers. Small shops line the road with a barber, a tailor, a butcher and several markets carrying most anything. The town was built by the Shell Oil Company in the early 1940s as a base for oil exploration. When they did not find adequate reserves to merit further work, they moved on, leaving an airstrip and a number of buildings. The Ecuadorian army took over some of the buildings and the GMU bought a dormitory, dining room and kitchen and several residences. The GMU property, a three or four acre compound was surrounded by a fence. Missionaries traveling from the jungles made it an overnight stop on their way out of and back into their jungle stations. The property had a Bible school for Ecuadorian students as well.

Our house stood at the back property line between the campus and an airstrip. Only a drainage ditch separated our home from the airstrip behind the compound. Annual rainfall for Shell Mera averages 250 inches per year. Though the soil is very porous, the airstrip needed good drainage to get the water off the strip quickly. Over the years a common purple orchid began to cover the sides of the ditch. Our kitchen/dining area looked out on those orchids. Now and then I went out early and picked a few orchid spikes and put them on the breakfast table. That was something we didn't have in Arizona, much less the United States.

The HCJB Shell Mera hospital was just down the road some 200-300 yards. Beyond HCJB, stood the hangers for Mission Aviation Fellowship (MAF) which had once been the home base for pilot Nate Saint. With other missionaries coming and going through the MAF's facilities and the hospital, there was always something to distract our focus on language study. For me this became an exercise in self-discipline. The town of Shell Mera extended a half mile along the road before reaching the gate to the campus.

The Ecuadorian Army planes, commercial planes, private aircraft and MAF planes all used the airstrip. There was a lot of traffic and Garth was always enthralled with the airplanes and their activity. The airstrip was a frontier airstrip where most anything could and did happen.

The Ecuadorian Spanish people can be quite fatalistic. This was demonstrated one day as pilot Dave Osterhus and I visited the airport hangar. We walked the half-mile distance and across the airstrip to talk to the pilot of a commercial plane parked there. It was an old well-maintained DC-3. A mechanic was fueling the plane for a trip out to a jungle airstrip. Fuel was leaking from the tank onto the ground near a man standing under the wing. As he waited for eventual boarding instructions, he smoked a cigarette while holding a three-pound can of black gunpowder under one arm. Calling Dave's attention to the situation, we agreed that the fellow was pushing the envelope of belief in life being what it will be. We finished our business and hurried away.

Over the next ten months we devoted our time to Spanish language studies with Malcom and Mardelle Brown, excellent teachers in the Bible school. As part of our language study, we conversed with a Bible school student for an hour each afternoon, helping us with our conversational language. We were asked to help out with maintenance of the school as our studies permitted. Maxine home-schooled Garth and during the afternoons he enjoyed playing with the married students' children in the afternoons. On weekends we visited various small groups of believers up and down the Shell road.

On Sunday afternoons Maxine and I sometimes walked over to the Pastaza River, wondering at its tremendous power. The Pastaza Gorge had a 300 foot waterfall about halfway up the gorge, bordering the road. To get through the pass at that point the road builders had to tunnel through rock some 40-50 yards to make a passage. With time and the abrasive action of the river water, the falls retreated up-stream several inches per year. The road from the pass was cut into the mountainsides, leaving a one-lane road that often was closed due to landslides. Passing another vehicle was a bit harrowing. A landslide would halt all through-traffic stopped until the road was cleared. Until the slide was cleared, buses took passengers from the jungles up to the slide at which point they disembarked and crawled over the slide to a waiting bus on the other side of the slide area.

Primitive farming was practiced on the steep slopes of the mountain. Seen from a distance, the mountainsides were check-marked with small plots of crops where farmers had cleared the jungle by machete to plant their crops. In this particular area of the Andes, they grew a fruit called naranjilla that grows on a bush four to five feet tall and about as wide. The naranjilla is a small orange colored fruit about the size of a tangerine. The fruit is shipped to the cities of Ecuador where it is peeled and juiced for a morning drink with a wonderfully unique taste. The land was so steep it seemed the only place in the world where a farmer could fall off his land and kill himself. The farmers had to work so hard for their poverty level life-style.

The culture of those living along the road and the mountainsides was fascinating to see. In some areas deep in the jungles of the Andes mountainsides, far from transportation and communication, families inter-married to the extent that some became mentally and physically diminished. Most were illiterate and had little knowledge of the outside world.

One man, native to that area and literate at that, asked Malcom Brown, "Señor Brown, the water of the Pastaza River, there is so much that flows year after year. Where does it go?" As Malcom tried to explain, the man just shook his head, mumbling that it was impossible to understand.

The Browns took Maxine and I on a day-trip to the area out along the road from Shell Mera to Puyo and then on along the Napo River road. We stopped at a small gathering of huts, meeting people who had moved from one of the isolated mountain villages to try their hand at farming in the lower lands. Listening and observing, we became aware that they were inbred, short small people whose thought and speech processes were poor yet they made a living on their farms. As we stopped, a man working his crops recognized the Browns. He called to someone unseen, that person in turn hollered to someone further yet in the distance. Within a few minutes 15 or 20 people had gathered asking Malcom and Mardelle if they would hold a church service for them. They were eager to take a few minutes from their work to enjoy a time of worship. We all gathered in a small building ten feet by twenty feet with one door and no windows. Malcom prayed with them, spoke a few minutes from the Scriptures offering encouragement and exhortation. As we left, those few people were so happy and wished us all well.

We stopped at another house to see people who were friends of the Browns. They were so happy to welcome us into their house. We visited a few minutes and they insisted, due to the local rules of etiquette, that we have a glass of milk. There were maybe two glasses in the house for a family of maybe five or six. They were a young couple born and raised on the frontiers of the land and socially uneducated. They discussed between them who should be served first: a husband and wife, the two men or maybe the two women. They finally decided that the two men, Malcom and I were to be served first, then Mardelle and Maxine.

Malcom and I drank our milk and handed the glasses back to the lady of the home. She washed and dried the two glasses under filthy conditions. We knew the risk of parasites and bacteria but as missionaries of all people, we could not refuse. There was no running water in the house. The well was quite a way off so they used the same water several times in order to conserve it.

A week or two later Maxine fell very ill and was

diagnosed with a severe case of the amoeba parasite, from which it took months to recover. Her strength and health went down hill fast. At the hospital, the doctors had to kill all the bacteria in her digestive system to get at the amoeba. Then, they began to feed her so as to bring back the healthy bacteria minus the amoeba. Little by little over the period of a month she regained her health and some of her strength. Eventually she was allowed to come home. In all probability that one glass of milk which she drank infected her.

Before we finished our language studies, we made a trip to Macuma to get some idea of what lay before us. The flight over the jungles and landing on the small grass strip was like nothing we had ever done. As far as the eye could see, there was nothing but jungle, broken by narrow strips of rivers or streams and an occasional Indian house. Approaching a large clearing, we saw Macuma for the first time. Closing in quickly on the community, we circled to get an aerial view before the wheels touched down on the grassy dirt strip.

After meeting members of the community and seeing the various aspects of the ministry, we talked with Frank and Marie Drown about the history of Macuma and its people. We had come to Macuma for a specific purpose and we discussed our role in what might be accomplished as we made plans and set goals. My role was to nurture, water and cultivate while providing a climate within which Frank could more easily evangelize, disciple and plant a church. God would make our efforts grow. I felt comfortable in that role believing this was God-given. In hindsight I am convinced the Lord provided just what we could handle, nothing more, nothing less.

The work in Macuma was divided into the following areas of ministries:
- Evangelism and church planting.
- Evangelism/Discipling ministry through radio broadcasts.
- Medical clinic work.
- Agricultural and community development.
- Station management and maintenance.

I would be responsible for the last two ministries. In a day or so, we returned to our home in Shell Mera to complete our language studies. I am not a linguist and tend to blunder my way through. Maxine was much more inclined to get it right but both of us found the Spanish studies interesting. We enjoyed the learning process.

When we applied to become missionaries, both GMU and the Alliance Academy were very concerned about whether Garth would receive the educational opportunities he required to be successful in life due to cerebral palsy. He had been affected both mentally and physically. At that time, the school did not offer a special education program for those with disabilities. We agreed that when the time came that the Alliance Academy could no longer help Garth learn, we would return to the United States to seek further education for him. Our time in Ecuador would be over. We understood that we only had a few years to work in Macuma and so we chose to forego learning the Shuar language. Many of the Shuar spoke Spanish and we could communicate by speaking Spanish, picking up Shuar vocabulary as we lived with the tribe.

We had finished our language studies but had not yet begun to pack our household goods, when I received an early morning radio message from Frank Drown asking if I could take an afternoon flight to come out and help with a mechanical problem that was quite urgent. We had not yet received official permission from the Field Council to move to Macuma. Thrilled to get out of Shell Mera, I kind of *jumped the traces,* catching an afternoon flight.

John Keenan in the small Piper Cub plane flew me with my tools out to Macuma. All planes needed to be out of the air and safely on the ground by about 6:00 as it got dark quickly in the jungles. After unloading my gear, I stood aside to watch John taxi to the end of the airstrip. As I watched the little plane rise against the dark background of the Cutucu Mountains to the west, I marveled at the unique and beautiful scene. I felt a sense of aloneness, wondering what the future really held for us.

Moving all our belongings in a Cessna 180 was no small chore; it took a good number of flights to accomplish the move. We would frequently pack boxes into the plane, fly out to Macuma, dump the contents of the boxes on the floor of our new home, then return on the flight to Shell Mera. Without time to sort and clean, one mother mouse found an up-ended book a cozy home for her babies. The pages were reduced to shreds, padding a little nest. In time we cleaned, and sorted our belongings. We established a living arrangement with the tarantulas in the attic. This was the home that Barbara Youdarian had built after the death of her husband, Roger. Now the five split-story building standing on the edge of the Macuma river valley would become our habitation. After two and half years, we were finally *home.*

In August of 1963, before we could settle into our home and dig into the job ahead of us, the missionary radio station HCJB asked the GMU for the loan of my time to help them construct five radio towers. Some months before we arrived in Ecuador, a terrific wind of hurricane force had taken down five of the radio towers at their antenna farm thirty miles from Quito. The loss of the towers reduced the ministry of HCJB by quite an extent, especially in the medium or tropical wave band and the AM wave band.

The HCJB engineers had learned that I had been a welder before coming to the field and in that the two missions had been working together on a GMU radio station in Colta and another in Macuma, they felt at ease in asking for my services for maybe a month. Because our ministry in Macuma had not really begun yet, the GMU field counsel asked me if I wanted to do the job and if so they would approve the project.

So Maxine and I, having just moved our household goods to Macuma, left things unpacked. We put together what little we would need and took the month-long job in Pifo, Ecuador.

For me it was a good ministry, for Maxine it was a bit lonely. Other women on the campus were busy with their own lives of ministry leaving Maxine alone to figure out how to deal with her time.

The engineers had designed the towers; we built them to that design. We built the towers in 20-foot sections to be bolted together and then raised into place as one tower. Four of the towers were 250 feet high and one 330 feet. They were not stand-alone but guyed with cables.

I set up a master jig to hold parts in place while they were tack-welded into place. HCJB hired local men to cut all the pieces needed by hand and another welder to do the final welding. Though I was fully capable of welding the final passes, my expertise lay in tacking things together so that all the pieces fit correctly and that the section was square and parallel.

We got them built and bolted together and then raised in their proper places. My job was finished. Maxine and I packed up and went back to Macuma to resume our ministry there. A year or two later, I did one other job for HCJB while we were in Ecuador: that was to overhaul one of their smaller diesel generators. I had some diesel experience but not nearly enough though I completed the job. Over the next few years I was called on to work on small, one- to five-kilowatt generators at small mission stations in the mountains and jungles. Some were successful, some were not.

We finally settled into our home, filling the cabinets with our possessions and sleeping in our own beds. After taking time to analyze what was there and what needed to be done, the work I saw ahead was to say the least daunting and overwhelming. The Ecuadorian government had given the GMU about 650 acres to build the farm, airstrip, school and clinic facilities along with residences for missionaries.

As part of our efforts, we were to teach the Shuar people how they could leave the nomadic life of hunting and gathering to settle into a permanent agrarian culture. The old hunter/gatherer culture of the Shuar was no longer viable. The whole culture was breaking down. The Shuar were becoming impoverished and their population was diminishing.

Tribal legends tell of living in the foothills of the Andes Mountains before the Spanish came into the country. As the Spanish encroached upon their territory, they fled to the lower elevations and the eastern jungles, displacing the Achuar people who fled further to the east ahead of the Shuar. Geographically, they could flee no further. We were there to teach them how to survive alongside the white culture and to teach them about the one who loved them, Jesus Christ.

Survival meant teaching them to raise cattle and to enhance their garden plots with more productive plantings. The income gained from sale of cattle would provide income to pay for medical treatment, schools, and the household things they once made from jungle resources. To merely give them these services and things would have been a dead-end trail.

We needed a farm to raise food for the school children that were boarding on the station. We needed pasture for cattle to raise the calves that would be given to the communities to start their own herds. The farm would serve as an example of the agriculture we were teaching them. When we arrived, the development of the farm had just begun with much work and a long road ahead.

My other duties also included maintaining the airstrip and arranging flight schedules, providing and maintaining the electrical generating machinery and the water system for the station. As the school day finished, I would work with the schoolboys. I also helped maintain the broadcast radio tower.

In the early days of the project, I bought the Shuar cattle, butchered them and flew the carcasses to buyers in Shell Mera. The teaching and training process involved hiking out to the various communities to help and advise them about disease control, quality of animal care and insuring the quality of the pastures. I taught the Shuar to raise and market cattle. Along with working in the communities, on many occasions it fell to me to mediate disputes between adjacent land owners. For many years, family groups lived far apart, in fear of revenge in the blood feuds that had been extended down through the generations. Now they needed to learn how to live in closer proximity.

We learned that many times having a vision does not include a program of continued maintenance of what has been built and completed. In the jungles maintenance is a terrific problem due to the constant deterioration by termites and moisture/humidity. With the task of raising funds for each project, one is inclined to do a job to the point of using it, not sufficiently to call it complete. Arriving in Macuma, we found much unfinished business.

Agricultural development was to be my primary ministry; community development came second and all else came third. At times, this plan became irrelevant in the face of all that need to be accomplished. We had agreed that Frank Drown would work primarily on the spiritual and church planting, and then the radio and schools. Maxine and Marie Drown would help in the schools and the Indian store, ordering and receiving supplies from outside. Marie helped with translation and language. The school administration was taken over by an Ecuadorian male teacher who worked out marvelously. Later on, nurse practitioner, Sarah Watkins, took care of the clinic. Frank and I were both responsible for cutting trees, dragging them to the sawmill to be cut up into lumber properly stored until used for the station's building projects. Toward the end of our term, Frank and I shared responsibility for constructing and maintaining the hydroelectric plant.

I went to work with the farm, providing bananas and plantains, rice, peanuts, yucca and pinto beans for the school children. Due to the mud during the rainy season, I began to think about building a more manageable cattle-husbandry facility but frequently found that I was working through the perpetual break-downs of the station systems. When I found a little spare time, I worked to provide a safer and more user-friendly airstrip.

I found two young men who seemed apt at working with cattle and developing pastures. Two or three other young men wanted to work with the farm equipment and I taught all of them their jobs. They retained their own homes in small communities near the station and I rotated the cowboys with two or three other young men in the community. The young men I used with the farm machinery were unmarried and so they stayed with me for longer periods of time. Eventually, I was able to hire an Ecuadorian Spanish man to oversee the work of the cattle and I depended on him to lead the Shuar cowboys in their work. Without these men to help me it would have been impossible to accomplish the work.

I began the project of fencing all the pastures and areas where cattle would be pastured. We planned to rotate two herds through the pastures and began cutting the jungle down to plant more grass. With five or six fenced pastures I could give the grass time to rest and re-grow after having been eaten down. The system worked quite well.

We did try some exotic pasture grasses in an attempt to find the best and most efficient grasses. One was called elephant grass and resembled bamboo. In the early stages it showed great promise but after a year it grew so rank and heavy that I regretted the day I planted it. Other grasses were wonderful but only worked well in small areas of several acres.

As the sawmill produced more lumber, we built a large milking barn. We brought in the cows morning and evening to both milk them for our use and to give their calves access to them. A second barn housed the calves. The two barns helped greatly but eventually became unusable. The vampire bats found the calves and made nightly raids on them. One evening, I stepped into the calf barn and a quick movement drew my attention. After I called out, one of my cowboys sheepishly stepped from behind a supporting beam, holding a short blow-gun. No self-respecting Shuar would be caught using a blow-gun made for the tourist trade.

"What are you doing?" I asked.

"Don Donaldo, does not this blow-gun move around the corners much easier than my blow-gun at home?"

I started laughing and after a moment he laughed with me. I learned the cowboys frequently did battle against the bats with darts and blow-guns in the shadowy confines of the barn, seeking to protect the young calves.

With the rainy season, the holding corral for the milk cows became bogged down with mud. The cows could only barely move through it to their milking stalls. This made for a very difficult situation.

Farmers in the mid-west had given us a small grain combine powered by its own engine and pulled by a tractor. We had three tractors, a three-bottom plow, a disk, several trailers of one sort or another and a flail-type mower for cutting the airstrip. I bought a welder that was on its own trailer as well. We used a World War II vintage jeep for hauling and transportation around the station. All of that needed to be kept under cover against the weather so we built a large shed and shop with workbenches and storage areas. This was a wonderful help to us, moving the equipment out of the rain.

The first year we lived in Macuma, a young boy attempted to steal a little kerosene from the generator shack. Dropping his lantern, the fumes from the barrel of fuel ignited and within minutes, the shack was engulfed in flames. The men on the station raced to the shack and began tearing the sides and thatched roof from the building, leaving less fuel for the flames.

The families gathered in the Drown's home a short distance from the fire, wondering if the flames would spread. One person suggested forming a water line from the spring. Grabbing what buckets were available, they lined up, widely spaced on the hillside between the spring and the fire. It quickly became apparent that they were too widely spaced to maintain a steady stream of buckets flowing against the kerosene-fed fire. Ruth remembers standing alone in the dark with an intense sense of helplessness as their efforts petered out. The silhouettes of the men stood dark against the fire's glare.

The station had no fire fighting equipment and if the flames spread, the other buildings would be incinerated. Late into the night, the men watched as the fire burned down, intensely thankful that the flames were contained to the one building.

After the fire, we built a more permanent housing for our three generators. For a year following the fire, we could look up at the branches above the pile of ashes and blackened stubs of lumber, remembering that night. In time, we built a shed for fuel storage and another where we did our butchering. For the latter, we poured a concrete slab with a hoist over-head. After killing the animals, the hoist made short work of skinning and quartering each cow. All in all, we had some good facilities. We lacked only a refrigeration unit to store butchered carcasses. The unit was obtained after we left in 1969.

Permeating all my work and efforts, whether in the shop, out in the pastures, orchards or fields, or on the trails, was the constant opportunity of counseling, persuading and convincing men of the things of God. My work in agriculture gave me a point of contact to talk and converse, to live my relationship with the Lord.

Midst the building of the farm, Frank Drown built a radio-house for the broadcast transmitter, studio and repair shop. He built two schools with classrooms: one for the boys and one for the girls with their respective dormitories. Eldon and Phyllis Yoder arrived several years after we did and built a residence near our home.

The school ministry included three Ecuadorian women teachers who served also as dorm parents for the girls and three Ecuadorian men teachers, one of whom was an administrator. The boys pretty well took care of themselves outside school hours. The culture required twenty-four hour protection for the girls.

To further understand the Macuma station I should

tell you something of the radio outreach to the Shuar people. Through the shortwave broadcast, the radio ministry reached a radius of maybe 50 miles, taking in some 15 communities and family groups. We were often surprised when a Shuar traveler passed through Macuma telling us the location of his home as we had not known our broadcast signal was getting that far.

Our agricultural and community development projects were reaching out to some seven or eight communities. We had given calves to these communities, helping them start cattle herds in each community. Our Bilingual teacher/school efforts were reaching out to some five or six of the same communities. Our total outreach was significant and wide spread, even reaching into the white population areas of Sucua, Taisha and the southern jungles.

With each passing month, we learned the ebb and flow of the jungle around us. We watched the river fog creep noiselessly up the Macuma River gorge. We marveled at the rapid development of an oncoming storm. Other days we enjoyed moderate temperatures with soft breezes in quiet meditation. The moods of the jungle were constantly changing. Some were rare, others a common daily phenomena—all were fascinating. The jungle is a complex unit of God's creation that is continually changing—destroying or rebuilding—but never in quite the same manner as before.

On May 10, 1963, a beautiful day was drawing to a pleasant, quiet close. The school children had left for the weekend. An old Shuar and his wife were approaching the clinic, the last customers of the day. The old man had his hand covering his eye, his head bowed with apparent pain in his eyes. His wife was leading him by the arm at a hurried gait toward the haven that has cured so many ills and complaints of the people: our clinic. I called for the Spanish girl who had for the present taken over the nursing duties of Marie Drown. She came and was opening the clinic door, at the same time telling the old couple to go to the front of the building.

At this moment, 5:20 p.m., the earth began to shake and jerk with a violence that I had never felt before. I spread my feet apart a little to maintain my balance and called across the little canyon to where our home is located, for Maxine to get out of the house. From that spot, I watched five buildings shudder as if a giant, unseen hand was trying to shake them apart. I fully expected to see them collapse under such treatment. Trees about us waved as if they were a whip in the same, unseen hand. In the few earthquakes that I have been through, I have never experienced one so strong as this one.

The Spanish people, being very gregarious, ran to gather in a group, fearing but not exhibiting it—fearing what, they knew not. The few Shuar finishing business of the day became very afraid. The old Shuar man and his wife ran to the corner of the building and squatted down with their hands over their heads. Other Indians came running, talking excitedly as they came.

The main quake deteriorated into strong frequent tremors at intervals of fifteen to thirty minutes. Standing by the clinic, I had a good view of the falling jars of pills, liquid medicines and powders. The noise amplified the disaster in that building. The Drown house, as I surveyed the damage later, suffered the least of any of the buildings. Our house suffered the worst. The kitchen floor was a mess. The refrigerator door had been thrown open, allowing its contents to flow freely on to the floor including milk, juice, eggs and other foods. My library was in complete chaos. The big bookcase had fallen to the floor with its contents and the smaller one had been emptied somewhat of its books - all were on the floor with paper and envelopes, paper clips, pencils, odds and ends. Later, I found that the clinic building had been moved on its foundation blocks, some as much as an inch and a half and all cement foundation blocks were either cracked badly or completely broken down. Our main power plant suffered no damage. Only the barrel of cooling water had tipped over.

Strong tremors continued to come and go throughout the night. Rain, thunder and lightning moved in during the early evening. The thunder from above and the rumbling of the earth below in its rearrangements were the same in sound so that one never knew when there would be another tremor. The night was filled with terror and fear for the Indians.

Morning came slowly, but as it has for so many eons, it did come to the jungle. The day dawned bright and sunny—tremors continued to keep us wondering if there would be another quake of equal intensity of the afternoon before. We continued to feel the sounds and movement of the earth, knowing it was not thunder for there were no thunder clouds in the vicinity. An occasional tremor after these dark noises did not let us forget.

From the Shuarnum came tales of fear, of falling trees, cracks in the earth, or river banks falling in, and thoughts of God and His judgment. Some tales are true while others are fantasies of the fearful mind. There came tales of many people accepting the ways of God and turning from the ways of the devil whom they had been following. We waited for the fear to subside to see what would come from these decisions made during such a time of emotional stress.

Sunday came two days later, a beautiful day, made all the more wondrous by the reactions and responses of the people. The church attendance was about 250. The previous Sunday the attendance was about 160. There were people in church that I have never seen before. I assumed they were men and women who were afraid and wanted to get right with the Lord. Such was the case. At the end of the sermon that was quite appropriate, men and women ran to the prayer rooms without urging. Some decided to walk in the ways of God, while others who had strayed, wanted to get right with the Lord and asked forgiveness for their waywardness. Sad, yet wonderful, that God has to show His power in such a manner in order to bring them to Himself.

The following Wednesday, God's presence was still being manifest in the jungle. As usual we had our weekly morning prayer meetings. Usually Frank Drown would take charge of the men Marie Drown taking the women. But that day the Drowns were in Quito and unable to witness the working of the Lord in the hearts of the Shuar people.

There has been a feeling since Sunday that something was going to happen among the people. The previous night we had asked the people on the farm to pray for the services because we understood we were going to witness God's presence among us. Again, there had been people around the farm that I have never seen before. People came with baskets of food as if they were going to stay for a day or so and they did. More people came, some were from far away.

That morning people began to gather early about 8:00. They wanted to get this repentance started. By 9:00 they were asking why we had not started the meeting. The men met in one place, the women in another. I took the men's meeting and a national girl took the women. Several of the men had asked me what caused earthquakes and what they were, so I decided to say a few words about what they were and how they are caused. First, we had some singing and then testimonies. Testimonies by men who have not testified for quite some time, by men who wanted to come back to the Lord.

While we were singing, I saw the older school boys come toward the building where we were meeting. They asked their teacher if they could all attend as they wanted to get right with the Lord. This had never happened before.

It is very hard to explain what an earthquake is to men who have no basic knowledge of the earth itself. Yet, it was a blessing to tell them because it brought back to my mind the basics of the earth and how the Lord made it. I tried to tell them that the Lord was not angry with them but only wanted them to think about Him and to follow Him. He loves them and wants them to love Him, but they will not. He has to exhibit His power to them and this is one way in which He demands their attention. They listened to every word that I said through an interpreter as I was speaking in Spanish. What a pleasure and an experience. After I finished, several expressed the desire to go over to the church building with the women so that they could have a prayer meeting over there. Some wanted to accept the Lord as Savior and some wanted to get right with the Lord. We all went to the church. There again, the Lord did wonders in the hearts of men and women.

Some of the Indians came to tell me how they fell on their knees to confess their disobedience to God and to ask God to care for them. Others told me they went to other believers in the vicinity to ask how they could know God and get right with Him. And so God used a natural phenomenon to speak to the Shuar people here in the jungle.

Out of this earthquake would come a spiritual renewal of men of God for the purpose of speaking to the Shuar people. Some spiritual growth would be made fast and glorious only to die just as quickly. Others would grow slowly to reach solid spiritual stature on which those less mature will be able to lean and depend on for help in time of need. There is no complexity in the jungle with these people—they were simple in their needs and desires. Their education and understanding was simple but sometimes this simplicity has tremendous depth and eternal value.

During the first 18 months of our time in Ecuador, through mother's and dad's letters, I knew that Pop's health had begun to deteriorate. Six or eight months before he died they were concerned that he would not last long and he did not. For our family, long-distance phone usage was scant, certainly they did not call to a foreign country.

A couple of weeks after the earthquake, I received a letter telling me that Pop had passed away with congestive heart disease. Though he died of heart disease, his lungs were filled with asthma and arthritis in most of the joints of his body caused him constant pain. As I read the letter, I was sad. I closed the door to my office, sat alone and thought for a while about the good times we had, about the fact that we would not sit together and talk about this and that on this earth.

My grandfather, Pop, was a close friend. He and I had many good times together. He taught me much. We communicated well. He never wrote me with his own hand, he told grandma what to write and say, and the message came through. He certainly could write; on the rare occasion, he wrote to his mother. I suppose with a fourth grade education he felt that writing was not his gift.

Pop was not a man to grieve long and hard, I would not do so either. I knew that another mile marker in our lives had been passed. In those days a death in the family at home was not cause for a trip home to mourn with others. The financial cost of such a trip relative to our income made it prohibitive and such a trip never entered my mind.

Along with her other responsibilities, Maxine homeschooled Garth for the first two years in Macuma. After the two years of home schooling and consultations with the Alliance Academy in Quito, we decided it was time to send him to the school in Quito for his third year. He would live in the GMU Dormitory where Ruth lived.

After the first year there, the Academy asked us to have him tested in the US to learn more about how to teach him. We made the arrangements for the trip. Maxine, Ruth and Garth took the bus to Quito enroute to the US while I remained in Macuma. As they picked up their papers, they learned we had forgotten one paper which surely only the government of Ecuador would require: Written permission from me allowing my wife to leave the country without me. In the last ten years, the world has learned that there are a number of countries requiring the same sort of paperwork for departing wives. When Maxine asked me to send such a letter, she used the MAF jungle radio network. All missionaries listened to the radio as it is a connection to the outside world. That day there was laughter and familial banter in the homes of many missionaries. I put the letter on the next plane up and after receiving it, they took the bus to Guayaquil.

With tickets in hand, ready to board the plane, the authorities pulled her out of line. They would not be allowed to board that flight as she had been bumped by the authority of an Argentine military general. And that was life in a Latin American country in those days. Nothing unusual there. The airline apologized to Maxine and gave her a room for three in a fine hotel in Guayaquil, taxi fare, the evening meal and breakfast the next day. She cried a little in frustration knowing that my father was to meet her and there was no way to inform him that she would not be there. Dad figured out the reason for her 'no show' and returned the next day. Three weeks later, they returned without a blip in the arrangements.

In the early 1960s, because of the uprising by the Mau Mau people in Kenya, Africa, a good number of the British and European farmers in Kenya felt they needed to move out of the country to a safer environment. Many were wealthy enough to take their farming operations to other countries and some moved to Latin American. Those growing coffee and tea chose the mid-altitude jungles of Ecuador, Peru, Bolivia and others.

There was a German couple about thirty minutes drive from Puyo that I wanted to visit. They were growing coffee and beginning to grow tea. Ecuador had only recently begun to produce and market tea. I drove out for a visit and they welcomed me warmly. They showed me around their operation, making suggestions for Macuma.

Late morning brought a break in their daily routine. They asked me to stay and have coffee with them. They poured me a cup and then passed me a pitcher of milk. I turned down the milk, suggesting that I preferred my coffee black. They looked at each other and chuckled. Right then I figured I was in trouble. I took a sip of that wonderfully fragranced coffee and found myself with a coffee that was so strong and bitter. I smiled and asked for the milk. They laughed and passed the milk.

There was also another new farm out in the jungles over which MAF flew over when they made a trip to Macuma. It was a massive project in which the jungle was chopped down and then without burning it, new coffee plants were planted. To one side of that project a smaller project was underway in which only tea was grown and processed. I decided to visit both farms to see what ideas they might have that I could use.

I made arrangements to borrow a vehicle from the Bible school in Shell Mera and drove down. I had to cross the Pastaza River on a hand-driven ferry that carried one vehicle at a time. I made the crossing in the morning without mishap, drove to and visited the farms. Late that afternoon I left intending to cross the river and get back to the Bible school. When I arrived at the river there were several vehicles lined up but the ferry was tied up for the night. The river had risen during the day to such a level and ferocity that the ferry operators felt it was unsafe to cross. They told us to come back tomorrow morning. So I drove back to the smaller of the two farms, owned and operated by a young Scotch couple. I told them my predicament and asked if I might stay the night. They welcomed me in and soon welcomed another fellow who apparently was a friend. We spent the evening talking and drinking tea. I drank more tea that evening than I had ever consumed at one sitting in my life.

The next morning, I left after breakfast and drove to the river. The Pastaza had gone down to something of normal level, becoming much more cross-able. When the people at the Bible school realized I had not returned they asked if the river was up. 'Yes, it was very high.' They knew I had not been able to make the return crossing.

I am a person with an almost insatiable curiosity to know the 'what and why' of almost everything around me. While in Macuma and Ecuador, the breadth of the spectrum of interests open to me was awesome. There were all aspects of Christian evangelization and church planting through the ministry of radio broadcasting, working with the schools and person-to-person contacts. All sorts of social work and many aspects of anthropology were open. Economic development and of course my chosen field of agricultural development was a big open door. Within those fields were the specialized fields. All of them could be used to develop a climate for and provide a point of contact for evangelization and building of the church among the Shuar people. As we approached a time to return to the United State for a furlough after five years in the field, I began to ask what could I do while in the US to enhance my ministry in Ecuador.

Before leaving on furlough some of the USAID technicians in Ecuador gave me written introductions to people in the Department of Agriculture in Washington, D.C. I wanted to learn if there was a better manner of working with the iron-laden and porous lateritic soils of Ecuador and the jungles especially. So our trip home included a trip through Washington, D.C. where I visited the Department of Agriculture while Maxine and the children visited various sights in the city. Though I talked to several well-known experts in the field they could not tell me anything that I didn't already know. It was an interesting interlude but vastly unproductive in answering my questions.

Then I then found a training program for the VISTA program (Volunteers In Service To America). So I associated myself with them as they worked on the Indian Reservations of Arizona. After all was said and done, I found the year had been a definite waste of time. What I missed most of all was someone to guide me, to give me some idea as to what I ought to be doing to prepare for whatever lay before me. My efforts left me frustrated as I searched for someone wise with whom I could exchange ideas.

During our furlough, we lived in Glendale, allowing Garth to receive the most medical attention possible. He saw the doctors who had worked with him previously and Garth profited from the time there. We were able to work in the church that gave us a good deal of financial support and renewed friendships. The time to return to our home in Ecuador could not have come soon enough and we eagerly boarded the return flight.

Chapter 6

EVERYDAY LIFE IN ECUADOR

We returned from our time in the United States ready to resume our duties at Macuma. Our children returned to life in the boarding home. As I remember those last two years, many stories come to mind from our lives in Macuma. Allow me to share a few with you without the discipline of placing each along a timeline.

...Searching for a downed plane...

As I mentioned earlier, the station of Macuma was located thirty minutes by plane from the edge of civilization. The
pilots and airplanes of Missionary Aviation Fellowship became our lifeline to the outside world.

Airplane crashes in the jungles were not uncommon and it seemed that we learned of one every 12-18 months. It was one of those events you hoped you were never involved in directly or indirectly. There were several groups flying in the jungles: MAF, Wycliffe Translators, Ecuadorian Air Force and at least two privately owned companies. There was the large civilian/military airstrip in Shell Mera; a small strip in both Sucua and Taisha and a military base to the east of Macuma. There were three or four major strips in the eastern jungles of the country. Several times over the period of our time in Macuma, a small military plane landed on the Macuma strip bringing in a government official for whatever reason.

One morning on our daily morning contact with MAF, they told us that a military Cessna had gone down somewhere in a 15 to 20 mile radius of Macuma and that they, along with another plane or two, would be in the air looking for it. Would anyone at Macuma like to go up with them as part of the search party? Frank Drown and I offered to do so.

That afternoon on our Macuma broadcast radio, the news of the downed plane was broadcast requesting that if any of the listeners knew anything about the matter, to please get in contact with us. So not only were there airplanes flying looking for the downed plane, Indians on the ground were searching as well.

We flew for a couple of hours up and down the length of the Cutucu Mountains and in a search pattern over the jungles. We continued until we began to get sick and quit for the day. After a week of searching by plane, we found no evidence of a downed plane.

Some three or four weeks later, a couple of Shuar Indians arrived in Macuma to tell us that they had found the plane with all four passengers dead. From their description, the plane had nosed straight down into the jungles and all four passengers died instantly. The military sent soldiers in to recover what they could of the bodies and what was left. The bodies were so decayed and crushed up together that only one body bag was required to carry out what was recoverable. As one of the men said, "the mourners needed a few bones and the hair off their heads over which to mourn." And that is about all they got.

Thoughts of the possibility of our own demise in

such a manner lived in the back of our minds as we, via the efforts of MAF, flew in and out for our various reasons and needs.

...Incident at Ayui...

The day had started with a low cloud ceiling. I made the early-morning radio contact with Mission Aviation Fellowship in Shell Mera. Pilot Dave Osterhus was to go to Yaapi to pick up passengers and take them out to Shell Mera. Shell and Yaapi were open along with several strips in the south jungle. Well, I thought, no flights for us in Macuma that day. I had no radio traffic so checked in and closed down for the morning. We ate breakfast and I went out to work.

I worked with my two cowboys in the pastures for a couple of hours building fence and caring for the cattle. By mid-morning the weather had gotten no better with low clouds, foggy-sort of stuff. I was glad there were no flights scheduled. It was a dreary day indeed.

Mid-morning I had come in from the pastures for a cup of coffee and a few minutes break. I turned on the radio to see what entertaining radio traffic there might be, not my usual fare for the day but mere curiosity, I guess.

"Macuma, this is AGC 27. Do you read me?" Strange, I thought, early morning traffic had indicated a trip toward Sucua; he should be on the return trip to Shell.

"Macuma here—read you, Dave. Go ahead, what you got?" I responded lightly.

Then suddenly I was slammed back into reality, "Don, I've been trying to raise you for twenty minutes.

I'm circling Ayui; been over here for fifteen minutes and I have fifteen minutes gas left. Are you closed in there?"

"No way of getting in here, Dave. Clouds down in the trees, we're completely shut down."

"Don, every landing strip in the jungle has closed in. I have Ella Ficke and Nettie Buhler* with me. If Macuma doesn't open up in ten minutes all I can do is go into the trees here (crash-land), at least you know where we are. I have no other choice. This is an old Shell Oil strip with young trees on it, at least young trees give us a chance. Will you stay on the radio with us?"

Carol, Dave's wife broke in from the Shell Mera base: "Dave, the north jungle is opening up. Can you get in there?"

"Can't do it Carol, too far; you'd never find us in the jungle between here and there."

I paced the floor, praying, pleading, reasoning with the Lord for an opening in the clouds. "Lord, is this the way you want these three people to die? Lord, they could very well die in ten minutes, just five minutes flying-time away. If these three people die...open a way, God, please; they are our friends, Lord, your children, God. Just five minutes of clearing Lord, that's all we ask. Five minutes from there to here, Lord."

Some minutes later Dave was back on, "What do you think Don, anything open? Don, I have no other choice."

"Still closed here, Dave."

"Don, can you hear the plane at all? Ayui is only five minutes way."

After all these many years, exactly who was with Dave has become obscure but as I recall it was Ella and Nettie.

I stepped outside and if I had ever listened in my life, I listened at that moment. There was no sound of an airplane, only the sounds of the jungle in Macuma. I ran the three hundred yards to the airstrip, maybe I could hear better from out there. I heard only the pounding of blood in my ears. No sound of an airplane engine. I ran back to the house and as I came within hearing, I heard Dave on the radio, "Don, there's a corridor opening in the clouds toward Macuma. Are you open there?"

I took the five steps to the porch in one leap.

"Dave," I almost screamed into the mic, "it's opening over toward the Pastaza, toward Ayui. Clouds seem to be breaking up. Give it a try, Dave. Dave, come on in."

"Got it Don. We're on the way."

I ran to the airstrip, listening, watching. Low over the trees from the direction of Ayui the red and white Cessna 180 dropped in out of that corridor, Dave lowered the wing-flaps, throttled back a bit, the engine spit and sputtered, no time for circling to check the airstrip, they came straight in. The wheels touched the dirt, kicked up bits of grass and the plane coasted to a stop at center strip. The corridor through the clouds closed in again.

"God, Lord of all creation, thank you, oh my God."

We had company for lunch that day, Maxine fixed sandwiches for the five of us.

Late in the afternoon the jungle skies cleared, open weather from Macuma to the Andes Mountains. We poured the needed gasoline into the Cessna's tanks and Dave, with Ella and Sarah continued their flight to Shell Mera.

...*The Fight Between Tiviaram and Mangash*...

I wonder why it is that as one lives his life it takes so long to learn things that seem so obvious.

It seems so obvious that little sins beget bigger sins and they in turn beget bigger ones yet. If such sins are not taken care of quickly, there will be consequences one would just as well not have to deal with. I wonder why we don't learn that lesson early on. Indeed, we do learn it and then immediately forget it—until the next time.

As a leader and a missionary in such a community, one receives all the problems that arrive on the doorstep of that community. Without a sense of humor it would be impossible to maintain any sense of perspective. And so it was as I lived in Macuma.

Surely if one could watch the on-goings of any community anywhere in the world, one would have to agree that they all contain about as much comedy as tragedy. The daily on-going of a typical Shuar community is not much different from any other around the world. There is as much comedy as there is tragedy, with all levels of one and the other in-between the extremes.

I was going about the day's work; whatever it might have been that day, when Tiviaram arrived. He met me face-to-face in the middle of the road. I greeted him as I did all Shuar who arrived at the station and I asked what was going on, why they had come, where they might be going. I could generally expect a friendly greeting and open answers to my questions in return. But never before had I, nor did I ever afterward get such a response as I got from Tiviaram that morning. I had never seen him as angry as he was that morning; in fact I had not seen *any* Shuar Indian as angry as he was. Tiviaram was so angry he could not even talk; he could not say what he had come to tell me. He tried but all he could get out was a stuttered jumble of noises that resembled something between the Shuar and Spanish languages. No greeting, no answers, just a flabbergasted inability to tell me what he had to tell me. As the scene played itself out, I was close to laughing at the show he put on.

Very quickly other Indians gathered round us to hear the story, to learn the gossip of the community. Again Macuma was no different from any other community in that respect. The jungle grapevine in Macuma was alive and well.

It was evident that there was big trouble and I wasn't at all convinced I wanted to hear it because I would have to deal with it. But here I was and I was determined to deal with whatever it was step by step. In earlier times surely such anger would result in a murder, and then another one and then another one in years to come.

Little by little through fits and starts, the story came out.

Tiviaram owned a fifty-hectare farm maybe a mile beyond the east end of the airstrip. He was a spiritual leader in the church, a man of some wisdom and renown for making good decisions in the community. He was a prosperous farmer, better with cattle than crops. He was a powerfully built man, his wife Maria was a loving, quiet and intriguing woman; Tiviaram and his wife had two children. Maria maintained one of the cleanest homes in the neighborhood; both the parents and their children were clean and well-mannered people. On the day of this incident I learned for the first time that Tiviaram had a temper that could only barely be held within reasonable bounds.

Across the jungle road from the Tiviaram farm a man by the name of Mangash had his fifty-hectare farm. Mangash and his wife had three children. Mangash was certainly a man of contrast to Tiviaram in most any respect you want to mention. Mangash was a fair farmer, better with crops than with cattle. He was not as careful a farmer and it seemed that he always had some sort of problem preventing his success. Mangash and his wife were more playful people, seeming to have more fun in life than some. They were not as clean as the family across the road but found other things more important than cleanliness in spite of what I might have thought. On the day of this incident, I learned that Mangash was a man of his word and a man who took things into his own hands when the community justice didn't work.

Not long after Maxine and I arrived in Macuma the first time, a Justice of the Peace arrived to register and perform civil marriages for any Shuar couple that wanted that service. Mangash and his wife availed themselves of that service one afternoon. They felt it would benefit their three or four children.

The following morning Mangash and his wife were scheduled to meet the incoming flight. They were late. As they came trudging up the trail and into the middle of the airstrip with Mangash in front carrying nothing but his rifle, his wife came trudging along behind bent over under a basket of 60 or 70 pounds. With everyone watching and waiting for them, Frank Drown suggested that having been just married, they had slept late. Everyone present whooped and hollered at the joke. Mangash and his wife laughed as loudly as any of those gathered.

As a farmer, Mangash seemed to enjoy growing rice and was proud of his small crop grown from seed I had provided him. I had gone down to see it at his invitation. Mangash cleaned and burned the jungle before felling the trees. Then, he and his wife came along and cleared the trash for further burning and the land was planted to yucca, corn, sweet potatoes and 'papa-china'—and a small patch of rice. Impenetrable jungle-brush lined the roadside along his land. Such brush served as a kind of fence to keep people and cattle from his land, at least made it more difficult to enter. I say made it more difficult: when cattle want to get from here to there they will find a way it seems, brush or no brush, and so stray cattle had gotten into Mangash's patch of rice two or three times already. It so happened that his one-acre patch of rice was close to the road behind a log of some 40 inches diameter. The log lay across the path to their house and one had to climb over it to continue on the trail to the house. Each time he had closed off the entrances made by the cattle until the only place they could enter was to jump over the log on the trail to his home; his home being some one hundred yards back off the road.

Mangash had complained of such damage to his rice but the elders and leaders of the community hadn't paid any attention to him; they just seemed to ignore his complaints. He had had enough of it and was determined to do something about it and suffer the consequences. I must give Mangash credit: he was a long-suffering man.

One night he saw to it that all entrances through the brush were closed tight. Then he went into the jungle and cut eight or ten poles three or four feet long, two or three inches thick and sharpened one end very sharp. He took the poles and planted them on the backside of the log that lay across the trail. He set them at an angle so that when a cow jumped over the log to get into the rice patch she would conceivably be impaled on the poles. It was possible that such an animal would be killed if the sharpened pole entered the right spot. Unfortunately, Tiviaram's cows got loose and wandered that night and fortunately they did not attempt to enter Mangash's rice patch. As Tiviaram went looking for his cows the next morning he came across Mangash's defensive tactics and immediately knew what would have happened had his cows chosen to enter the rice patch that night.

Tiviaram was livid with rage and determined to spread the news of the terrible sins of Mangash. He seemed to overlook his own *sin* of being careless that his cows got loose to ravage other people's rice crops and bean patches. He seemed a bit unconcerned as to what might happen to his neighbor's crops—just a bit arrogant in the matter. He and Mangash were from different families. I am not sure if they were related any where in their ancestry. It was rather a miracle that they tolerated each other, living so close to each other.

Obviously I was the one to solve this problem: I was a gringo, I had brought in the cattle, the rice, the schools, Christianity and therefore I was culpable. Or it seemed that their thinking went thus.

I walked down to see the evidence, to talk to Mangash to get his side of the story and then make a judgment. I tried to convince Tiviaram that he was as much at fault as Mangash but that didn't go over very well. At least I prevented a killing and things cooled down for a while. The sharpened poles were removed and Tiviaram did a better job of tying his cattle.

The incident was not forgotten, however. Several months later, Mangash ambushed Tiviaram and beat him up pretty badly. His head was cut in several places requiring stitches. Tiviaram pleaded total innocence, parading his broken and bleeding head before any who would look and listen. Sara Watkins, our nurse-practitioner sewed up his wounds and that seemed to be the last of the incident that I knew of.

...Canoe Trip down the Cangaimi...

There had been sporadic communication with the Atshuara people, three or four days' walk to the south of Macuma. Atshuara men had come up from an area we called Putuimi, an area on the border between the Shuar and Atshuara people. We wanted to extend our work into that area with radios and an airstrip. We decided to make a canoe trip down the Cangaimi River to Putuimi.

The agenda for my participation in the trip was to help with the planned evangelistic effort on the trip, and to find out just what the rest of the jungle looked like, what was there, who was there and if I could extend my agricultural development ministry to the region. Frank Drown had his own agenda as did the doctor who went with us.

The planning went something as follows: A few of the community members from the Putuimi area had walked up to Macuma—a two- or three-day trip—and asked if it might be possible that we come down and preach to them, tell them more of Jesus Christ, to visit them. In such a situation I was always a bit cautious with my enthusiasm because many times that sort of plea is nothing more than a ruse to get us down there so they could request help with schools, cattle, airstrip, etc. The Putuimi community was situated on the Cangaimi River. It seemed that such a trip might be easier done by taking a canoe down the river rather than walking.

There was a community just south of us called Cangaimi situated up on the bluff above the river. Colon Altamirano and his wife were missionaries living there. There was an airstrip, co-ed elementary school, church and clinic all operated by the Altamiranos. We talked to Colon about buying a canoe from the Indians in the area, one large enough we could use to make the trip down the river. We ended up with one weighing about a thousand pounds that would carry ten men and their baggage.

We asked for a doctor from HCJB who might make the trip with us to minister to the medical needs of the people along the way in the places we stopped. Dr. Wally Swanson agreed to go with us bringing along another doctor visiting from the States. We also asked two or three Indians from Macuma and two or three from the Cangaimi area to go with us. It seemed to be a good working group.

When asking Indians from different areas to help as a group it is very important that one keep in mind the ancestry and history of the men and/or women involved. One must make sure that there has been no history of animosity among those who are to travel together. John Stuck told of an incident in which he traveled with two or three Shuar men. They were to stay overnight at a certain house. The men with him spent the whole day talking among themselves trying to learn/remember family relationships in order to make some family connection between themselves and the family in whose house they were to stay. They finally arrived at the conclusion that way back in their history they were indeed related and they felt they could easily stay at that house.

I remember that the day we were to leave we called in MAF to fly us from Macuma to Cangaimi and that took some time. I think we left the Cangaimi riverside about 10:00 traveling a few miles the first day. It took a bit of time to learn to work together as a team: running through rapids or over difficult passages.

The Cangaimi runs through a canyon until it gets out of the mountains and onto the flatlands. In the mountains the river ran a good number of rapids and it was not an easy job maneuvering to run them with such a canoe as we used.

Down in the flatter lands, one finds more bamboo growing along the riverbanks. The bamboo is of the sort that puts out runners with very heavy thorns or spikes at the nodes of the runners. The spikes are so strong they can easily puncture and tear one's flesh from the bone.

At one point, we started into a rapid without really checking it out before committing ourselves to running it. It was too late to turn back when we saw that at the bottom of the rapid the river ran into a rock cliff and then turned sharply to the right. At that point bamboo canes and runners bent over into the river in such a way that we would inevitably have to run through it. We ran down the river banging into the cliff and through the bamboo. Literally everyone in the canoe took to the water to avoid the bamboo. Fortunately the canoe was not damaged. It was a beautiful day full of sunshine, warmth and quiet beauty. The river water was so pleasant when we bailed out, it wasn't all that bad—I think that fact helped us through the unfortunate event at the rapids.

The trip further down river was beautiful. The forest rises tall and luxuriant at the river banks. Now and then we came across a small clearing with a house. Once or twice we stopped to rest and visit for a while.

We came to a place where the river had cut through a small range of steep and rugged hills. The cut was very narrow with walls on each side. In the passage, a five-foot waterfall required that we remove all the gear from the canoe and portage the gear around the falls. The rocks were slick from a constant mist and wetness, making it a dangerous and time-consuming operation. When we came back up the river a week or so later, getting back and over the falls was really difficult. I ended up on the other side of the river from the rest of the crew. To get back on their side they threw me a rope, the end of which I tied around my waist and I jumped in the water while they pulled me across. There was no way I could have swum across such a millrace.

As it descended from the mountain area of Cangaimi to the lower elevations, the river slowed considerably and became wider and more placid. The trees at water's edge were larger and hung further over the water. There was a greater danger of stepping on stingrays in the few rapids we did encounter there. The temperatures were much warmer and under the full sun we sunburned quickly.

We stopped one night at a house near the river. I am not sure whether or not they knew we were coming. Word could have traveled telling that we were on the way. At any rate they were pleased to see us and hear the gossip of the world about them. We unloaded the canoe and pulled it well up onto the riverbank. In the house I threw down my sleeping bag in an open spot where I thought would be convenient. The man of the house watched me and then spoke to Frank Drown. Frank then told me that I would have to move my bag as the spot I had chosen was the spot the people had buried a son some years earlier and they didn't want anyone sleeping on that spot. I did not learn the details of the son's death.

The next morning we rose early as we still had a day's travel to do. We were getting our own breakfast while several children in the house were running here and there and going outside. The man of the house was sitting on his favorite stool accepting the breakfast his wives brought him. Suddenly he sat up and hollered at one of his two wives: "Quick, bring my daughter back in the house before the devil gets her." The little girl had gone outside before her father had wandered out to see that all was well. The girl was brought in and the man continued with his alcoholic breakfast of chicha. I won't go into how chicha is made, but it is alcoholic.

Moments later, one member of our group looked over and pointing at my face, he started to laugh. Everyone looked and started laughing. A vampire bat had feasted on my blood during the night and left a spot of coagulated blood on the tip of my nose. I was the only one on the trip to provide such fun and frolic at the expense of my blood. Though the scar it left was not large, it was visible for six or eight years afterward. Indian babies and young children need to be covered at night to prevent such attacks. Indian mothers seem to have an instinct to know when bats are attacking their children and they awake quickly to drive them off.

Any warm blooded animal is subject to their nightly raids. Their exceedingly sharp incisor teeth make a quick incision and inject a liquid that is both an anesthetic and an anticoagulant until they finish feasting at the *fountain of one's blood*. They can drain the blood from a chicken in a couple of nights. They also tend to return to the same fountain as long as the body can be found in the same place each night.

After breakfast we gathered our gear, stowed it in the canoe, said our good-byes and were off. We reached Putuimi in mid to late afternoon and made our camp in a shelter the villagers had made for us. We would be staying there a few days so they deemed it more comfortable for all concerned to build us our own shelter. The river at that point was quite wide and sluggish; maybe fifty yards across.

Putuimi is a village maybe a half-day's walk from the area of the Atshuara Indian people. In seeking more contact with them, Frank Drown had announced over the radio the previous week that we would be in Putuimi and anyone interested should come up and visit us. And so they did. Several men with their respective wives and children arrived at the house to see who we were and what we had to say.

The men sat around a fire outside the house: eating

lunch, visiting, talking as people do the world over. The women and children sat behind listening and talking. Dr. Swanson noticed one of the women sitting apart from the group. She was quietly sobbing and crying. Dr. Swanson asked Frank what her problem might be. He in turn asked one of the men what the problem was; in that case it was improper to ask the woman directly. We were told that the woman had just been bitten by a conga ant. The ant is about two inches long and has a very painful bite with the pain lasting several hours.

Dr. Swanson said he thought he could help her, went to his bag for a syringe and medicine. He gave the woman a shot of an antihistamine. Within minutes the pain was gone. I shall never forget the wonder on the woman's face as she considered what had just happened: medicine that could stop such pain, a pain that all Shuar had had to endure for centuries.

As we traveled down the river to get to Putuimi, ten or fifteen miles before we arrived, the land had flattened out and the speed of the river slowed. Actually, the river became quite wide with small sandbars and grounded tree trunks and such trash. We began to see evidence that alligators inhabited the river and their presence was a subject of conversation as we traveled those last few miles to Putuimi.

That evening after we were well settled and had had our dinner, our Indians suggested that it might be fun to go alligator hunting. They suggested that the meat was good to eat and who knows what other adventures we might come upon. The river in front of the village was broad with several sand bars and grounded logs, good hiding places for alligators. Frank and I and three or four Indians took the bait so to speak.

It was dark and we had brought along both my .22 pistol and a good flashlight. Frank and I sat in the bow, the Indians with paddles were in the rear. Out on the river, not far from a large grounded tree trunk we heard a loud splash and figured the noise came from a large alligator. We slowly swung the light around to catch the yellow gleam of the alligator eyeballs—maybe one or two were just floating along waiting. We did catch sight of two spots of light and knew it signaled an alligator—how big it might be, we had no idea. Slowly and soundlessly the Indians propelled the canoe toward the animal. Frank held the light on the eyes in the river and I cocked my pistol. At about ten feet from the animal I aimed for the point between the eyes and pulled the trigger. The eyes disappeared. As the canoe glided by the place the animal had been the Indians reached down and grabbed the tail and one foot, swinging the alligator into the center of the canoe—it was about four feet in length and apparently dead. One in the canoe and we began to look for more.

Yes, the alligator was apparently dead and *only* apparently—suddenly it exploded into action, jaws gaping, and tail swinging violently from side to side. Two Indians jumped for the bow to land on top of Frank and I, and several more landed on the stern on top of whoever was guiding the canoe. Everyone was yelling, "Kill it, kill it!" Finally one of the Indians landed a good blow with a paddle to the animal's head slowing it down enough to be able to grab a leg and tail to swing it back into the river from which it came! Voila—end of alligator hunt, enough is enough. Everyone had a good long laugh at each other and ourselves as we paddled back to shore. It was good to get to bed for a night's rest.

The next morning Dr. Swanson thought to do a bit of fishing using the rod and reel he had brought along. I do not remember what he used for bait. Soon he pulled in a fish about eight inches long that was identified as the dreaded piranha fish. Not long before he went fishing he had used the river to brush and wash his teeth—we quickly found other sources of water to brush and wash our teeth. Alligators and piranha fish were sufficient to keep us out of the water.

We stayed a few days, preached and taught the Word of God The doctor treated one and all that arrived to plead his services and then it was time to get back upstream and home. Consideration of the journey home was no pleasant thought. We all knew that it would be nothing but work. On the trip down the river one could ride the rapids and even in the extreme sections we could get out and walk the canoe down through the rapid. On the return trip we would have to drag the canoe up through each and every rapid and that translated into work, hard work. That consideration was one of the reasons we took as many Shuar with us as we did because we needed their strength and help. We paid them well.

One afternoon we were moving along quite well in a stretch of deep water with hills rising from the riverbanks; we were able to motor along nicely. One of the Indians excitedly indicated an animal in the river quite a way up stream. We looked and there was a tapir wading along in the water. I had brought my .30-.30 carbine with me thinking that maybe something of this sort would show up. I pulled the rifle out and took a rather quick off-hand shot—hoping for a hit. I did hit the animal, wounding it. Within a minute or so we were at the point the animal had hit the bank. I grabbed my .22 pistol and followed the Indians. We caught up with the animal fifty yards up the hill. It was an easy shot to finish the kill.

The Indians carried it back to the canoe, dressed it, laid it in the canoe and we continued up river. The first house we came to we figured we should stop for the night. The meat had to be cared for, we intended cutting the meat into small pieces and smoking it—that was the best way to deal with it. There was about 150 pounds of meat, all of it good.

The Indians cut the meat into three or four inch cubes, then pushed a pointed stick through each piece to be able to turn it over the fire as necessary. They made a grill of sorts with wood stakes high enough over the fire to keep the stakes close enough that the meat would smoke and dry quickly without burning the stakes. Several of the men stayed up all night working with the meat.

We had to get the job done quickly both for the sake of the meat and to not prolong the trip. The Indians would have stretched the process out over the next 24 hours if they could have. The next morning we got up, packed the meat in baskets and were on our way with an extra 100 pounds or so in the canoe.

We had another two days on the river before we arrived at Cangaimi where we started. The trip up river had been a long hard one and everyone was very happy to get back home. I found it interesting to note the selfish possessiveness two or three of the Indians manifested over the meat we had brought - they wanted it all for themselves. We prevailed on them to give it to the school children in Cangaimi. I have no idea what the

Indians felt like after getting home and resting over the night. I was tired for a week, unable to do much physical labor. It had been a successful trip and I felt good about it. We had done some evangelizing work, encouraged some Christians and arranged to for the Atshuara build an airstrip in the Putuimi area.

We made a second trip six or eight months later to find the airstrip almost built and ready to use. A month or so after the second trip, MAF pilots were able to take an alternate route on foot to confirm the usability of the strip.

...a few days of vacation at the beach...

To a sizable degree I am a workaholic. I am very loyal to my job and to the work assigned me. Except when totally exhausted, vacations seemed to me to be something I could not afford time-wise or financially: there was just so much work to do and so much that needed to be accomplished. We did however, take several vacations and the following stories concern some of those trips.

Maxine loves to tell the story of a conversation with several other missionary couples who worked with us about my need for a vacation. I was speaking to someone telling him or her of my doubt that I needed a vacation. The person to whom I was speaking commented, "If you don't know you need a vacation, then certainly everyone else around here knows it. And furthermore, if you don't take one, everyone else will...en mass!" So I took the hint and we traveled to a reputedly nice Ecuadorian beach for a few days.

Our trip to the beach was taken during a national

spat between Ecuador and Peru. The United States had come down on the side of Peru and that did not bode well for Americans in Ecuador. I once saw a letter from someone in Peru to a recipient in Ecuador. The envelope had, of course, a Peruvian stamp on it. The Ecuadorian post office canceled the stamp with a solid black blob of ink obliterating the Peruvian postal stamp indicating Ecuador's proud rejection of everything Peruvian.

On our trip to the beach, we traveled via a public van-type bus. There were two passengers in the front seat with the driver. In back of the driver there were three rows of seats. Maxine and I and one other passenger took up the second row and were obviously the only Americans in the van. Behind us were two more rows of seats. On a long stretch between villages, one of the passengers in the front seat got into a very loud and excited discussion with a passenger in the seat behind us about the *war* with Peru. There was talk of atrocities by the Peruvians, how the Ecuadorians had turned back the Peruvian onslaught, and much more in very excited and animated voice.

As with natives in most countries I suppose, while in the security of their own land, people can be quite patriotic. The passengers knew we were North Americans and the United States was not at that moment in the good graces of the people. The next bit of discussion was specifically for our benefit. Maybe they intended that we go home and tell the president of the U.S. what he was up against. So the subject of the part that the U.S. was playing in the conflict came up. And the talk became more animated, more courageous until finally the man in back of us rose from his seat to proclaim that when the United States attempted to invade Ecuador, the men of the country would arise as one man to push the invading armies into the sea. And all the passengers on the van shouted their encouragement. Maxine and I chuckled and laughed. About that time the van arrived at another stop where several of the people disembarked and all was quiet from that point to our destination.

...the ferrocaril trip from Guayaquil to Quito...

The word *ferrocarril* is Spanish for railroad; the word, *tren* is the word for the train itself. In Ecuador there is a third term for a special vehicle that travels some of the Ecuadorian rails: the *autocarril*. It is a single-car vehicle made from a bus that has had railroad wheels installed on it and it travels the railroads of the country just as the trains do. We were told by friends that it is a *must do* trip at least once in one's mission career. And so we did.

The Ecuadorian railroad system was designed and built by the British in the early 1900s. The road from the coastal plains to Riobamba and Quito took a southern route. The geography of the Andes Mountains takes a gentler climb in the south than it does in the north and I imagine that is the reason for a southern ascent rather than a northern one. The climb was gentle all except for one place which was the infamous 'nariz del diablo', the devil's nose. It was a shoulder of the mountain, steep, high and extending toward the coastal plain: the road had to be built on that mountainside. The only way to do that was to build a series of switchbacks without curves: The operator would go-forward to a point, then throw a switch backing the train up the road to the next higher switch. Back and forth the little car rode through some five or six switches up the mountain. It was an awesome climb with all eyes focused on the mountainside we were ascending.

All four of us had taken the bus to Guayaquil. From there we spent time at the beach then returned to Guayaquil to catch the *autocarril* to Quito, reportedly a 12-hour trip. Early in the morning, maybe 5:00 or 6:00 we made our way to the train station and boarded the *autocarril*. Those who could not afford seats inside were given a spot on top of the vehicle, a common practice in the Ecuador of those days. Those who traveled the *autocarril* seemed to be a bit more sophisticated than those who regularly traveled the buses. Of course there were the ever-present *commerciantes*. These people were those who, with all they could carry on their backs, traveled from town to town, selling their wares: everything from foodstuffs to whatever could be carried and sold.

In the world over, towns and villages are settled. The railroad following as straight a track as possible, passes as close as possible to the existing towns and villages. And so it was in Ecuador. From Guayaquil we traveled south through miles and miles of banana orchards, pastures and crop land, climbing, climbing gently toward the Andes mountains. At every village and town vendors with hot cooked foodstuffs, of doubtful sanitary cleanliness, though typical of the place, day and time, besieged the passengers.

Climbing through the foothills the geography became more mountainous and the distance between towns and villages became greater. Toward the top the railroad passes through the *paramo*. This region above the tree line is covered with coarse clumps of grass. The

Quechua Indians, who were native to the land, harvested the grass for roofs on their huts, for fodder for their cattle and sheep and whatever other uses the grass might serve. The day was clear and now heading north through the high central valley corridor, we could see the snow covered peaks on both sides.

By the time we arrived in Quito, we were ready to arrive somewhere, anywhere. The gauge of the railroad in Ecuador is narrower than those of the US. The trains are slower and loads are lighter. Therefore the roadbeds are not quite as smooth and well kept. All that resulting in an uncomfortable ride that sways from side to side. Though it was a wonderful trip with so much new to see, hear and experience, we appreciated arriving at the station in Quito and taking a taxi to the GMU guest house for the night.

...Swimming the River at Night...

We had been in the jungles only a year or two living in Macuma. Ralph and Marian Stuck were taking the place of Frank and Marie Drown who had gone to the States for their one-year furlough. Ralph Stuck was the designated station head. I don't remember why, but Ralph and Marian were not at the station on this particular night. They had gone to Quito, possibly to visit their children and do shopping for themselves and the station.

The first revenge killing in over ten years had just taken place—sometime over the past 36 hours—and the community was pretty upset over the matter. It really didn't involve the population in the immediate area though some thought maybe it did. I was sure that some in the community knew who the killers were. Those who didn't know, wondered. I was not sure in which community it had taken place. I did know that it had not taken place in the community across the river. I was sure that it had happened to the south and over closer to the Pastaza River.

Revenge begets revenge, sooner or later, sometimes much later. And the neighborhood wondered if they would be next on the list of the vengeful. There probably was no family in the community in which somewhere in their history, revenge had not taken place.

The day had been a dark one. Fear hung over the jungle like the rain clouds over the Cutucu Mountains to the west. A sadness and anxiety, too, permeated the community. As evening closed in, Maxine and I sat at supper quietly talking about what had happened during the day and what could happen that night. We retired to the office to read and write before turning off the generator for the night.

About 9:00, we became aware of a sound from across the river, a crying, a wail and the beating of drums. We wondered if the men from the family of the dead Shuar had come to take their revenge. Shortly after the wailing began a young Indian man called from the front yard and I went to answer him. Anguash stood there alone. He was a relatively sophisticated young man who worked for us on the station. Anguash was from the Santu family that lived across the river and he said he was worried about what was going on over there hearing the wailing and crying.

The average Shuar does not like the night because they are taught from early childhood to fear it and to protect as they can against it. The night is the time the evil spirits roam the jungles doing the work of their nature. The Shuar will not go out into the night without the companionship of several others, and then only for an evening's hunting or an evening's killing. They need a good deal of encouragement even then. It was no small thing for Anguash to be in my front yard asking my help.

Anguash had come to ask if I would go with him to his house across the river to see what the trouble was.

We talked a minute or two. I felt that maybe I should go with him, though I certainly did not know what I could do in any event. As a man reputed to be a missionary, a man of God, I would somehow be able to make a difference. I told him I would go and I went to change clothes and get my shoes on while he should go down to the river and bring the canoe to our side. I said that if it was on the other side he should swim over and bring it back. I apparently did not know what I was asking of this young man raised on the lore and fears of the jungle. Anguash left and I went in the house to do as I had said. I brought along a flashlight as it was a very dark night.

When I arrived at the river bank some few minutes later, I found Anguash standing there and the canoe on the other side of the river. I asked why he had not swum over to bring it back. Anguash looked down at the river-bank sand, then looked up at me and said, "There might be a *pangi* (river boa) out there?" A number of thoughts went through my mind all at one time: principally, "...so, what is going to protect me from the *pangi* that won't protect you...?"

I pulled off my shoes and pants and laid them on the beach, put the flashlight on top of my pants, walked into the river and began to swim. There is no dark so dark as that at the river's edge in a jungle canyon with a cloud-cover. I could not see the other side of the river, indeed, I could not see my hand in front of my face.

As I walked into the water, I could feel that the river was at least a foot into flood-stage. I also knew that at that stage trees could be floating down river and I could not see them. Not being able to see I could only hope I would not encounter one. There would be snakes in the branches and I could get entangled in the limbs. I knew there was no *pangi* out there—at least I talked myself into knowing there were none. But I did know that there were river rapids just down-river sixty yards or so. Several Shuar had drowned in them in broad daylight so what chance did I have in them at night? I felt the current stronger than usual and swam as powerfully as I knew how, swimming up-river slightly to make up for the distance the current was carrying me down-stream. I swam as I never had before or since. I was so relieved when I banged my foot against the rock and gravel on the far side of the river.

The canoe was pulled high on to the riverbank. I dragged it into the water and called to Anguash to turn on the flashlight to give me a point to guide to and pushed off. We returned to the far side and made our way to the few houses from which the wailing was coming. Anguash talked to the men of the house, they told him the women were very much afraid. The women had just started wailing out of fear. I never learned if their fear was justified or not. I did not feel it was. Certainly there was a seriousness in the air that did not let go for some days. A head-shrinking was a serious thing and everyone seemed to know that there was trouble yet ahead. Pastaza Province where we lived was under the jurisdiction of the Ecuadorian Army and they would see that justice was carried out.

Anguash and I returned home without incident. He thanked me for going with him. The trip seemed to relieve the tension he felt as well as among those who came to the farm the next day.

The swim across the river was indeed one I shall never forget. The thoughts that went through my mind as I swam are now gone. But I can tell you that I thought of everything imaginable that could have gone wrong that would have brought my early demise.

...with the monkey's help...

Saturday's breakfast over, early morning chores were done and I settled into the chair at my desk. Then, a quiet call from the front yard. "Don Donaldo?"

The mere decision to sit down and spend the whole day relaxing with my books and typewriter, just reading and writing, had somehow been sufficient to lift a large burden of responsibility from my mind, for one day at least. Maybe it was selfishness, maybe it was bone-deep fatigue, I really do not know. And then that quiet call from the front yard.

I got up slowly, stiff and tired from the activity of the week just ended, begrudging the intrusion. From the front porch, I looked down at the young Indian who had called and greeted him, "Good morning, Tsuambra; what do you want?"

"Good morning, Don Donaldo," he responded. "I don't want anything; I came just for a little visit."

Just a little visit, I thought sarcastically. Tsuambra

never came for a visit and I was sure he was not here now for that simple purpose. I wondered what he wanted. One day a week I want to be left alone and he, of all people, has come for a little visit. Tsuambra was one of those young men I did not choose as a close friend. I don't know why, I just didn't like him. We didn't fit, I reasoned.

I sat down on the step and waited for the next movement of Shuar visitation protocol. I did not intend to prolong the process, so I waited in silence. Shuar Indian men are always uneasy in the midst of silence and I used it as a weapon.

"Nice day, isn't it?" he commented, looking toward the river.

I looked out across the canyon and confirmed, "Yes, it is a good day." And I sat there. Tsuambra looked over toward the store across the ravine from my house. "Are you going over to the store?"

"No. I have no reason to go," I said. I sat quietly.

Tsuambra nervously shifted his weight from one bare foot to the other. He looked from the store to the river
canyon and then a furtive glance at me. I still did not know why he had come and I did not intend to ask.

The night before, I had sat down to supper with my wife. I had bowed my head and thanked God for the food and that the week was over and I was so glad. I had spent the whole of it tramping through the jungles. Day after day, I had been called out to visit and talk with Indians about their cattle, their chickens, their dogs, and I was glad they didn't have any cats. I encouraged them, I treated their sick cattle, and I listened to their family and inter-family problems. There had been cattle that would die in spite of my medicine, and there were inter-family fights ready to explode into murderous ambush. They told me of this one or that one who said he was 'God's Indian' but really was not. I had decided that today I would stay home and rest. I was going to relax from the frustrations of so few answers, so few real solutions to so many problems and troubles.

Saturday in the eastern Ecuadorian jungles, is the day that people come to the local store to buy their needs and their wants. Aluminum cooking pots, cloth for dresses and shirts, an axe head, gun powder and shot for muzzle-loading guns, and who knows what else. They, like most people in rural and developing societies of the world, I suppose, like to make a day of such an outing, especially if they have had to walk an hour or two to get to the store. They go to the store first, always the store first, then, depending on the urgency and amount of money left from their purchasing, they go to the clinic for medicine for the family. If need be, they look for me to sell them medicine for their cattle and dogs. During this day-long process, there are many people to greet, and gossip to learn; thus the day is used until it is spent and time to start for home in the late afternoon.

As I sat on the porch, I desperately hoped that Tsuambra did not need medicines for his cattle. If I sold him medicines, a dozen more people would also want to buy medicines and my morning would be shot. Who knows what the afternoon would bring. I just wanted him to go away. I did not want to talk to him or anyone else. I wanted to get back to my books, to my typewriter, to myself.

Tsuambra could stand my silence no more. "Bueno,

Don Donaldo, I am going to the store." With that he turned and left.

I responded, "Bueno, I'll see you later." I got up from the porch step and went back to my desk.

The day passed quietly and without further interruption. About four that afternoon, I got up from a nap and settled down in my study again, with a book and a cup of coffee. By courtesy of the old Wallensak tape recorder, I listened to music by Chopin. Just slightly over the sound of the music I caught the noise of the monkey's chain slapping the side of the house as he played there. That noise told me a visitor had come. I put my coffee and book aside, shut off the Wallensak and went to see who was playing with the monkey.

Anguash, the monkey, was often the center of attention when Indians came to visit. Occasionally, he had been the catalyst that eased a sometimes abrasive contact between two vastly different cultures, two vastly different modes of thinking and communication. The natural comedian in him calmed the mutual inhibitions of two cultures, making it possible for them to deal successfully with the need at hand, that of the gringo missionary as well as the Shuar. The noise of his chain slapping the house as he played was, many times, the signal telling us quietly that someone was out front wanting to visit. Whether we wished it that way or not, it seemed that the mood with which we answered the call to the front yard was dictated by the method used to let us know someone was there. When an Indian came calling us by name, we answered with some seriousness. When they came and played with the monkey before calling us, our answer was certain to be more relaxed and friendly. Anguash always seemed to be able to put things in a congenial mood for anyone who bothered to ask his help.

At four o'clock that afternoon Tsuambra chose to ask the monkey's help. He had not called so I didn't know who was there playing with the monkey. His return told me immediately that something out of the ordinary was about to be brought to my attention. I greeted him and asked a question or two about who had been to the store, who was doing what these days. He responded with light answers as he continued playing with the monkey. I was playing for time. Suddenly, he pushed the monkey aside. The laughter on his face changed instantly to an expression of dark intensity. My gut told me that I did not want to hear what was about to come out of his mouth. My day of quiet peace and relaxation was about to end abruptly, and I knew it instinctively.

"Don Donaldo, my cow is trying to give birth to her calf and she hasn't been able to do so yet."

I asked how long she had been in labor. "Since last evening," he responded, shifting nervously from one foot to the other.

I felt anger and agitation build within me. Why couldn't they leave me alone for just one day.

"Why didn't you tell me this morning when you were here? All day long you have been at the store doing nothing but talk, and now you want me to go help your cow," I responded.

His expression continued dark and in bitter despair he said, "Don Donaldo, you were not in a good mood this morning, you did not want to listen to my problem."

Suddenly I was very tired. I was sick in the pit of

my stomach. His accusation was true. He had come this morning to ask my help for his cow but did not because he sensed my mood. He read me like a book though he couldn't read a word on a printed page, not in his language or the official language of his country. He did not know what to do for his cow; he had come confident I would know and go with him to do what he could not. He waited hour after hour, all day long, knowing that his cow might die during the waiting and that if she did he would probably be blamed for her death. Finally, he could wait no longer, he must find out what his waiting had gained.

He stood there, deliberately waiting for me to answer. As he did so, I saw more in Tsuambra than I had ever seen before. He was not just a happy-go-lucky, work-when-he-had-to, wimpy sort of man. He was a man who had waited a long time to get what he needed, who had used the friendly playfulness of the monkey as a filter through which he hoped I would see and supply his need.

I apologized to him. I told him I was sorry about the way I had acted and I asked him to forgive me. He responded only by asking again if I would go and help his cow. I grabbed my bag of medicines and tools and we took off at a jog. His pastures were a good thirty minutes away, even at a fast trot. The afternoon had an hour and a half of daylight left.

When we arrived at his pastures, I found the cow was down and in a lot of discomfort. We had to pull the calf. It was dead. The heifer was far too young and small. She never should have been bred. After cleaning her out and working with her, we were able to get her up and standing. Tsuambra brought water and put feed before her, showing me that he did take care of his cattle when he knew how. I told him what more he should do and to call me if she went down again. We talked a bit, using the incident as a lesson on the matter of too-early breeding. I cleaned my tools and put them in the bag. He thanked me and I took the trail home.

I had done all I knew to do and there was nothing more I could do. I walked slower now there was no hurry. I waved to an Indian out tying his cattle for the night. I listened to the jungles, to the pastures, their sounds were quiet and peaceful; they were relaxing sounds. They were sounds I needed to hear, and somehow they put an end to the day, to the week.

The dusk of the evening was deep by the time I reached home. I stopped at the foot of the porch steps to say good night to my friend, Anguash, the monkey. I thanked him for his help that afternoon. From his windowsill-perch he opened one eye, looked down to see that it was only me. He curled his tail back over his head to continue the night in sleep.

...the time of the dog sickness...

Dogs are an important ingredient of the Shuar household. They are primarily hunting dogs and that is the primary direction of their training and usefulness. Without them, putting meat in the family cooking pot would be very difficult. Additionally, they are pets and watchdogs for the family home. With the dogs around the home there is not much chance of anyone approaching a Shuar house unnoticed.

They serve another very interesting purpose as catalysts in the relationship between husband and wife. The husband handles the family finances, the wife seldom has any money of her own not given her by her husband. When the need for or desirability of a hunting dog arises, or maybe he just happens on a good deal, it is the husband that does the buying. He may pay a ridiculously large sum for a good dog, because sometimes money means so little to the Shuar people. He takes the dog home and gives it to his wife and it becomes hers. In that he has bought it and given it to her, it is an indication to her that he cares for her and she understands it that way. Now the ball is in her court. How well she takes care of the dog is a good indication of how much she cares for her husband. Good hunting dogs in a family are a credit to the husband for buying them and knowing how to use them, they are a credit to his wife as she keeps them in good shape.

In September or October of 1964, a disease came through the jungles the Indians called the *dog disease*. It came through every four or five years in varying degrees of intensity. This year the disease ravaged the dog population taking at least 25 percent of them. When it hit a home with more than one dog, they all went. None was immune.

I had seen a few dogs that were sick, dogs that Indians had brought in hopes I could do something for them. I had sent medicines out to some, those I thought might help—they didn't. I tried other medicines and they only prolonged the death process. I had finally given up knowing that in reality I could do nothing for them. I think the people knew I had done all I could. I wished they would not bring their sick dogs to me: I found it stressful and raised only false hopes for their dogs. I knew I did not have veterinarian training. All

I could do was to read and try first one medicine and then another, all of my efforts amounted to a shot in the dark. I had talked to Ecuadorian veterinarians and they hadn't a clue concerning the illness. So in reality, all I could do was to say I was sorry. I could only hope our two dogs would not get whatever the disease was.

Thursday was an interesting day of the week—if I had had time to look at it that way, it was a very busy day for me. Thursday was Prayer-Meeting Day, the prayer meeting itself took up an hour or two of the morning. It was usually well attended by the Christians of the immediate community and surrounding communities were represented by a few who came for various reasons. For those who would rather not have had to deal with prayer meeting, it was the price they had to pay to get what they came for - whether that was to buy medicines, get medical attention by the nurse, buy consumer goods at the store, meet with prospective business contacts or just enjoy good conversation and the latest jungle gossip - it all happened on Thursday mornings at Macuma. Midst all the activity of the morning, eventually those who had need got around to bringing me news of their sick cow or sick dog: whatever they thought might interest me—and benefit them.

Pauch came bringing four young dogs. She was 40-45 years old, on the upper limits of being desirable as a marriageable woman. At that age if she lost her husband, she didn't have much hope for another good one. She was well dressed for a Shuar woman, wearing a clean light-colored dress. Her husband took good care of her. She carried the smallest dog in her arms, each of the other three she led on a leash of jungle vine.

From the porch of our house I saw her coming

and I already knew the scenario that would take place. I dreaded it. She too, must have dreaded it but was willing to pay the price of rejection and disappointment, she must have known. The dogs beside her trotted along so delicately, their movements were so quick, so light. Heavy mucus ran from their eyes and noses, they were very sick dogs.

I stepped off the porch steps to greet her. At some twenty feet from me she stopped and stood looking at me, then at her dogs wondering if it was all right for her to be there.

"Good morning," I greeted her, knowing it was not a good morning for her. I was not linguist enough to partake in the proper protocol of such an encounter. She knew it, all the Shuar knew it and graciously forgave me my abruptness.

"What do you want?" I asked, knowing full well what it was she wanted. I hoped maybe she would not ask for medicine for her dogs.

"My dogs, are they not very sick? Do you not have medicine that would make them well?"

"No, Pauch, I don't have medicine for your dogs. I do not know what to call this sickness, so how can I have medicine to cure it?" I said. I cannot tell you how much I wanted medicine for this sickness. Forgive the cliche, but my heart ached for this woman.

She took a deep breath and moved slowly to sit down on the porch steps. Drawing the dogs close to her, she began to cry quietly. She seemed to have come to the end of her road and it was a dead-end. Lord, how many times I had stood watching these people come to a dead-end and I had to stand there watching with no answers. All I could offer was sympathy for something I knew nothing about.

It bothered me a great deal that I didn't know what this disease was. Distemper, most likely, or something akin to it. Nothing could be done for distemper. I thought I had come to accept the fact that I just could not solve all the problems this jungle threw at me. But as I watched Pauch crying and I felt so deeply sorry for her, I was not sure anymore that I had.

As she sat sobbing, she asked, anyone who might be listening, "Will my husband want me now? My dogs will die. He will not have dogs to hunt with. Will he not send me away?"

Pauch lived several hours walk to the southwest. I did not know her husband. Several weeks later I learned she lost three of her four dogs. Her husband didn't send her away; he bought more dogs. I never saw her again.

The days and the weeks came and went, the disease continued to ravage the dogs in Shuarnum, the Shuar communities. The Indians quit bringing their sick dogs to me. In a month or two it had finished its work and went away. The Indians seemed to give a sigh of relief. One or two of them let me know that the Indians did not blame me for not being able to cure their dogs. "Are you God that you can cure a disease that has no name?" Always a question; all their words of wisdom were framed as questions, and they were at times wise beyond their knowledge.

It was always so frustrating to do all that I knew to do yet not accomplish anything of the desired end. I never knew but if there might have been just one more thing to try that would have accomplished that result, just one more medication, but what? How far must one travel an obvious dead-end road to show that there really is nothing more that one can do.

At times, the jungle was a fiendishly frustrating enemy, and at other times such a friend. I had a limited number of tools, only so much knowledge, and certainly no more strength than the average man—seldom did any of these ever seem to be enough.

When one sees so much to be done, so little time and never enough strength to do the job, one must ask the questions: where are the lines of priority, what is the important thing and what is it that can wait?

...Toxoplasmosis...

Along with the *dog disease* and our concerns with that problem, we noted a couple of other bothersome incidents that drew our attention. Our minds were definitely on medical things at the time.

Three prominent young men, each one from a different area, came down with a disease that blinded them. It wasn't a sudden blindness but came on over a period of several months. At the same time the dogs in the area showed symptoms of a disease we had not seen before. The doctors out at the hospital in Shell Mera examined them and hadn't the slightest idea what the problem was or how to treat it.

The matter bothered me both because of the loss to the young men and that we might have another *dog disease* on our hands, whatever that might be. I began to talk to the Shuar who knew anything about what was happening, asking for description of home conditions, symptoms of illness in their dogs, did they know of any such similar historical incidents. I worked up a brief of common symptoms and conditions then went to my veterinary books to look for something that might fit the notes I had taken.

One evening a couple of weeks later, I was perusing my books one more time and came across a disease description that seemed to fit perfectly: Toxoplasmosis. A day or so later, I went out to the hospital with all my notes and the book that referred to the disease and talked to the doctor most acquainted with the situation. He agreed with me on my diagnosis and began to examine and treat the young men for the disease. Their cases symptoms began to improve. I do not know if they gained their sight back. There wasn't much we could do for the dogs. The problem seemed to disappear and that was the end of it. I found the diagnosis to be a fulfilling experience after all my frustration.

...Pancho Gonzales...

When I entered Macuma, agricultural and economic development projects were presented to me. I made some general plans for the farm, including what I wanted it to look like in five years, how I wanted it to operate and what I wanted it to accomplish within itself as well as in the community. I intended that one day I would be able to say I had finished, close out the project and go back home. My purpose was certainly never to build a farm, an agricultural complex that would last beyond my days. I suppose that if someone had come to carry it on beyond our stay, I would have been comfortable with that.

I had been in Macuma almost a year and the cattle project was going well. The herd was growing, it's health was good and I had distributed a number of young cattle to the Shuar communities. Clearing the jungle and planting pasture was going well. The crops on the farm were also progressing, providing beans, rice, yucca, bananas and plantains for the boarding-school students, as well as some experimentation with new and hybrid crops. The growth and good health of the project over-all, to say nothing of my constant fatigue, indicated I definitely needed help to run and operate the farm, especially the cattle division, it was getting beyond my physical strength and the time available to deal with well.

Frank Drown came to me one day remarking that he had received a letter from an old friend named Pancho Gonzales. The gist of the letter was that Pancho was asking if perhaps he might be allowed to come to Macuma and work on the farm tending the cattle. Several years previously he had worked for Frank around the station, caring for a few cattle and doing other chores as were needed. I don't know why he left except that at the time the jungle station was a lonely place for a single Spanish man.

Frank told me of the man's abilities and qualities, as well as his problem with alcohol. We agreed that he would be a good man to help with the work on the cattle project. I felt too, that it would be something of a ministry to the man as he attempted to live out his relationship with the Lord.

Pancho was a Believer who battled alcoholism. He was in fact fleeing to employment in Macuma where no alcohol was allowed. Certainly with more Spanish people on the station, his social life would be easier than previously. And so Frank invited Pancho to come and work with us.

At the early morning radio contact a week or so later, Dave Osterhus announced Pancho had arrived and was ready to fly out to the station that day. We made arrangements for the flight to be made late morning.

After Pancho had climbed out of the airplane, claimed his suitcase, a cardboard box or two and put them into the airstrip storage shed, he declared himself ready to go to work. He claimed he would arrange his living facilities after the day's work was over, that his immediate responsibilities lay in accomplishing the work of the day I surely expected of him. In this manner the man began his loyalty to our relationship and me.

I had reserved a small building for him to live in, a one-room affair which had a bed, table and chairs. He added shelves and other items as he desired. He seemed quite comfortable in it.

Pancho Gonzalez was some 55 years of age, maybe. He was the quintessential jungle pioneer, a man as easy in the jungle as he was working a farm in a more populated rural area. He was a farmer and cattleman of experience. He was independent, capable at surviving. He was a restless man, that restlessness carried him all over the Ecuadorian jungles, from the north to the south. He blazed trails through the mountain jungles for roads years ahead of their building and in the process was twice lost for many days with only food and water from the jungle to keep him alive. He was married and openly claimed some five or six children of both holy and unholy matrimony. I doubt that his children really knew him as the father that brought them up. He was not a man famed far and wide for his jungle exploits but those who dealt in the taming of the jungles knew him.

I have often wondered how one's previous acquaintances effect a person's perception of present acquaintanceships and friendships. Pancho looked and acted so much like a man I knew while I was in the U.S. Navy, my Chief Engineman aboard the LST602. Their personalities were so much alike. I had a good relationship with the Chief Engineman and he was such a help to me, I have wondered how much of that relationship transferred to my relationship with Pancho.

There was no way I could have foreseen the blessing he was to be to me as a person, to the growth and ongoing work of the cattle project. I do not know what I would have accomplished without him. Pancho was a tall, lean man, by Ecuadorian Spanish standards. He stood maybe five-foot ten or eleven inches tall. Most Ecuadorian Spanish men stand some five-feet four inches to five-feet nine inches tall and carry a good bit of weight around with them. His slender build gave no indication of the considerable physical strength he possessed. When I met him, his head was covered with a large shock of black hair that had begun to gray. He always wore a hat made of plastic rather than straw or palm frond or the usual inferior grade of felt. I often wondered where he had obtained it. Pancho wore a mustache typical of the Ecuadorian Spanish male of his age. He was accustomed to shaving once a week.

Pancho was something of a cattleman, a man who knew how to make pasture and maintain it. He was a master at swinging a machete on the jungle growth. He knew how to care for cattle on jungle pastures. Pancho seemed to be able to do all and whatever he needed to do. Unfortunately Pancho's alcoholism limited severely his capacity to grow personally or spiritually.

Pancho's religion, his Christianity, his relationship with Jesus Christ was different from what I was used to knowing. I believe to this day that in spite of his singularity of personality, I will see him in heaven. He had been raised a Catholic. At some time in his life, middle age maybe, he became convinced somehow that the Catholic religion did not hold the ultimate answers to the questions that life threw at him. He encountered Jesus Christ through the ministry of a Protestant missionary in the Upano River Valley. He accepted the answers of Jesus Christ but just could not totally separate his relationship with Jesus Christ from the Catholic religion in which he had been brought up. When it came to knowing for sure how he was going to be forgiven the sins of his life, of his being, his faith was in Jesus Christ but most of his daily life was governed by Catholicism.

Pancho very quickly claimed his spot in the social structure on the farm. The Indian men seemed to take an immediate liking for him: they seemed to respect and work well with him. In the beginning I was a bit concerned, considering the usual attitude of the Ecuadorian Spanish toward the Indians. After watching him a few days, I gave him responsibility for the cattle and pastures, relying on him for advice as to how the cattle were doing and when they should be put into a different pasture. For the first month or two we talked every morning about the work for the day, what needed to be done, what I expected of him and what he thought should be done. It wasn't long before the need to talk about the work took place once a week and then it was his responsibility. He was a joy to work with.

Pancho was a loner. He seemed to work by himself even though there might be several Indian men by his side. There was never any great amount of conversation between him and the Indians around him, though he had the personal freedom to do so if he wished. He had a rhythm to his work that got a lot done. I enjoyed watching him work with a machete in the pastures as he knew how to swing a machete with great effectiveness.

Over a period of time and as opportunity was presented, I asked him about his family and life. I don't think I ever determined exactly his current marital status nor how many children he had. Pancho did not talk freely about his past or his family life. One day he came to me asking if he could invite a young son to come and live with him and work with him. When the boy arrived I was quite surprised at his young age—he was maybe ten or twelve years old. I did note that he was a bit slow mentally but certainly capable of taking care of himself.

The boy arrived and everything went well. I even asked that one of the Spanish teachers spend some time with the boy teaching him some of the basics of schooling though I think Pancho worked with him, teaching him to read. It worked somewhat, not real well.

He stayed with Pancho for a year or two and then Pancho sent him back out to stay with another family. I do not know what ever became of him.

...*The Blessing*...

One day I witnessed an event I have never forgotten. The incident brings to mind the stories of the patriarchs in the Bible. As the moment occurred on a dirt strip in Macuma, rather than among the tents of ancient Israel, I understood what I was seeing to be among last vestiges of a culture that soon would disappear from the world of Ecuadorian Spanish. The incident contained an act, which I had never seen carried out, only heard about. I feel privileged to have watched it.

Pancho had been on the station about a year when he came to me one morning and told me that an older son had written and asked if he might visit him. Pancho's request was merely a formality and show of respect in that there would seldom be a reason that such request would be denied.

A couple of weeks later the young man arrived on an MAF flight. Pancho introduced me to his son, Sergio. He was a quiet reserved man and stayed with Pancho at all times. I gave Pancho freedom to work or not work, as he so desired in order to spend as much time with his son as possible. He and Sergio enjoyed hours of conversation and companionship over a week or so of time in Macuma. Then it was time for Sergio to leave and he asked that he be allowed to leave on the next flight.

About mid-morning the MAF Cessna arrived with a load of cargo. While it was unloaded Pancho and Sergio walked slowly side-by-side out to the airstrip. I watched them, wondering about their thoughts and words at the moment. Dave, the pilot, had finished unloading and a bit of light conversation with Frank Drown our co-worker. It was time to leave. Sergio walked to the door of the plane with Pancho a step behind. He stopped, turned....and then it happened.

Sergio, a man of some 30 years of age dropped to his knees before his father, wrapped his arms around his father's knees and bowing his head, quietly pled his father's blessing upon him. "Father, bless me, bless me, oh Father."

Pancho quietly placed both hands on Sergio's head and mumbled words I could not understand. Sergio rose from his knees and without dusting off his white pants kissed his father on both cheeks, then silently climbed up into the plane.

Dave fastened the seat belts, closed the door and started the engine. Pancho stepped back and with hands at his sides, watched the plane taxi onto the airstrip and to the far end. As the plane turned and raced by us, lifting off into the sky, neither Pancho nor Sergio waved, they merely looked at each other, father and son. Surely something had happened here which at its depth I did not comprehend.

It is almost unbelievable to me today that I did not later ask Pancho about the incident: what he said and why and where he learned it. I knew that I had witnessed something special, something I would never see again. Surely, I witnessed again the end of an era, of a culture, of a way of religious thinking in another country.

As far as I can find out Pancho and Sergio never saw each other again.

Pancho was a wonderful man in so many ways; he was a sinner as we all are and so he had his faults and weaknesses. When we could sit down and talk, I enjoyed being with him.

For a period of several days Maxine was out of Macuma. I've forgotten the reason, maybe on a trip to Quito to visit the kids? One evening I was a bit lonely and I knew that Pancho and his young son were just sitting by the fire in their little cabin waiting for the time to go to bed. Seldom do such people get to enjoy such things as cookies, cake, and other desserts. Maxine had left a full cookie jar for me and I thought, why not take some cookies up to Pancho and his son and spend an hour conversing with them. I made a pot of coffee, put some cookies on a plate and set out for their cabin three or four minutes walk away. Pancho was so surprised, he began to laugh at me standing at his door with a pot of coffee in one hand and a plate of cookies in the other. I think I can safely venture the prediction that it was the first and only time in his life that a young gringo man had ever stood at his door with a pot of coffee and a plate of cookies suggesting an evening of conversation between the three of us. For months afterward Pancho might at any time begin to chuckle and remind me of that night I stood at his door with a pot of coffee and a plate of cookies in my hands waiting for him to open the door. We did talk long into the night about this and that of our lives.

The day came when Pancho finally informed me that he was leaving. He said he wanted to go back outside—I don't remember where. Sucua, the Coast maybe, I do remember it was hard to see him go. I would miss him on the farm with the cattle because it would bring me an untold increase in my workload. I was already working from daylight to dark, how could I work any longer hours? I would just have to do only those things I had time for. It was hard also because he had become a dear friend, a valued friend as friends go. He was a quiet rock of a man midst the storms that came and went on the farm. I missed him so much.

A number of years after he left Macuma and I had returned to the United States I inquired of a missionary friend as to his whereabouts and how he was doing. He told me that Pancho had moved out on to the Puyo-Napo Road with a son, I believe. Pancho died some 20 years later in a dark jungle hut of old age and pneumonia. I do not think he ever returned to the alcoholism that afflicted him prior to his coming to Macuma. I was thankful for that.

...Maashu Catani...

God brought another man, a Shuar, to work with me. Maashu Catani entered my life not because I wanted him to but because I had something he wanted. To get what he wanted he had to deal with me. I quickly found that I did not want to deal with him but he met my requirements and so I had to fulfill our verbal contract. The relationship started at rock-bottom and had no where to go but up and up it went. I do believe it was one of those relationships the Lord provided and guided for His glory.

At about age 17, Maashu had married an older woman for convenience. The woman was about 45 or 50 years old. Then when he had gathered enough money, he got rid of her and married a younger woman. While still living in his father's home, he had obtained a cow from his father to start his own herd. The cow injured her foot and she kept the wound agitated and open by constantly licking it. The only solution was to tie her so she could not lick it until the wound scabbed and quit itching. The only way to do that was to bring her to the Macuma farm and tie her in a stall. He would have to cut grass and feed and water the cow. He lived across the river, a 40-minute walk away. He complied with the rules for a week and then began to slack off. I accepted that it was a difficult chore and so helped out. The injury healed quite rapidly and he was soon able to take her back home.

Up until the time of the injury to his cow and his relationship to me, Maashu had been an angry young man. Life was not going his way and he didn't like it; he began to rethink the things about God he had learned from Shuar Christians and the missionaries. He and his younger brother, Anguash came to me one day and asked if they might work for me as 'cowboys.' I told them I would think about it, giving them an answer in a week or so. It was advantageous when I could hire two men from the same family.

I did eventually hire them. They moved their wives into the housing on the farm and stayed the usual three or four months. Two cousins or brothers then took their place while Maashu and Anguash and their families returned home to care for their land and gardens. They would return to Macuma at a later date. I hired Maashu to work for me several years over the six-year period I was there.

Six months after beginning to work on the station, Maashu, under the ministry of several people came to know the Lord very well. Sad to think of and mention, but years later and long after we left Macuma, he got away from the Lord and as far as I know is no longer walking with Him, and that to his and his family's detriment.

Maashu and I tramped the jungles together, visiting other Shuar and their cattle, advising them, counseling them and enjoying our time together. I taught him all I could teach him of animal husbandry during those times together. As the conversation would turn from one thing to another, we often talked of the Lord and how He wanted us to live our lives for Him.

...going with Maashu to treat a sick calf...

One week was going pretty well with work getting done and the people were all in good moods and happy. The weather was drier than wetter; therefore we were in the dry season. The dry season is usually warmer than usual. During this drier than usual season the jungle can be gorgeous. It is awesome to walk a jungle trail among the huge trees, trees with tendrils hanging from branches that are maybe 80 to100 feet up in the trees. I had done a lot of walking the first couple of days of that week and things in general were going well. I was always thankful for a calm week.

I found a good bit of enjoyment, even excitement in studying the cultural characteristics of the Shuar Indians. How they dealt with me, how they dealt with each other, how they approached a problem. When I asked a man to help me with the cattle, to live and work on the station, I had to be sure he was of the same extended family as one already working for me. A brother of one already working for me would have been about as good as one could ask for. The brothers, Maashu and Anguash came from a community across and up the river. Their father's name was Catani.

The trail to Maashu's house was beautiful with a lot of trees and vegetation of interest. I had only to cross the Macuma River and then it was pretty flat from there up river. There were no streams or other rivers to cross. The anticipation of walking the jungle trails was always a dilemma for me. Getting out on the trail for an hour or so took so much energy, I often wondered if it was worth it in the long run. At the same time, the joy of getting out on a nice day, seeing the huge trees and the plant communities that grow on them, to see the odd plants that grew on the ground, fungi that grew, matured and were destroyed all within a couple of hours was wonderful to behold. Setting my mind to the task, I could walk long and hard but by the time I returned home, I was tired. On a particularly warm day and on a return trip that crossed the river, it was always wonderfully delicious to plunge into the water to cool off, to wash the sweat from one's body—a reward from a trip well done.

Wednesday was the day of the week for the Macuma community mid-week prayer meeting and worship service. For ten or fifteen years I had been accustomed to working five days a week. Every day of the week was a work day and one worked all day long. That was the routine of my life and as far as I was concerned it was the best that one could do to accomplish what one had to accomplish in life. The people of the jungle saw things a bit differently: life need not be so organized with its routines set in concrete, so to speak. There was no such thing as time off, holidays or vacation. At the proper time they needed to go fishing, hunting or gather ripe wild fruit. Every activity is part of the day in the flow of life. Time and activity is linear: all of it important to another part of life. Fishing is certainly as important as clearing a patch of jungle for a garden. So what does one take time off from? For the Shuar to get up on Wednesday morning and come to the service was just as important as going hunting.

A different way of thinking about things and setting of priorities.

On Tuesday afternoon Mashu asked if he could go home after the meeting on Wednesday in order to treat his sick calf. He wasn't very clear as to what the problem was but it did not sound too serious. I told him that he should go and I gave him the medicines that he should use. On Wednesday late afternoon I noted he was still around the farm and asked why he had not gone yet. He said that well, maybe he should go on Thursday. I agreed with him. But again on Thursday he was at work. Now he would go on Friday. He did not go on Friday and I began to wonder what was going on. Was the calf sick or was there another problem he was not telling me about. When this sort of thing happens one has reason to search out the truth of the matter. Things are not always as they appear. I began to ask questions and it took only a minute or two before I hit pay-dirt.

Maashu asked if I might go with him to see his sick calf and help him treat the animal. I suddenly realized that he had been wanting me to go with him ever since he asked to go home. I don't know whether I was just dense or there was really no way of knowing. I have to admit to not being very talkative sometimes, but then in my own defense, people do have the responsibility of asking for whatever they want or need. I agreed to go with him though I was not too keen on it. Saturday mid-morning, Maashu, his bother Anguash and I got started.

I always reserved Saturdays to rest and do the little things in my office I needed to do. There was no doubt in my mind that I was not as strong as I had once been as I was getting tired more quickly. Or maybe I was just working harder and longer hours than I had the energy to fulfill. I was not happy to be on the trail and so I decided that if I was going to go, those boys were not going to amble along. They would walk my speed and my speed was fast.

We arrived at Maashu's house some forty-five minutes later and began to look for the cow and her calf. We must have looked for thirty minutes. The cow had broken loose from her rope, hiding the calf and we could not find it. Maashu, of course, was very embarrassed. I figured that if it was still alive three days after reportedly being sick, then there was no real urgency about the matter.

After thirty minutes, I decided to leave the medicines with Maashu and return home. Maashu insisted I go to his house and eat lunch before starting back. I knew this was the polite thing to do. I started to sit down on an empty stool by the fire but he insisted I sit on his bed, a more comfortable place. Suddenly people were doing all sorts of things according to orders Maashu was giving. The house was home to him and his immediate family, as well as for his father and mother and younger siblings. I soon had a cut and peeled papaya on a banana leaf ready to eat. Then I had a piece of boiled fish and yucca tuber also on a banana leaf. If I looked like I needed something else, I had it, whatever it might have been. I stayed in the house with them fifteen or twenty minutes, eating and visiting. It is not their custom to eat and visit at the same time as it is embarrassing for them to eat in front of another person. When in a group and eating, they stand with their backs to each other until finished. As I realized my curt treatment earlier of the two boys, I began to feel very humble at the treatment I was getting from them. Certainly this was more than I deserved. I had done nothing for them but walk to their house to treat a sick calf that they couldn't find.

I instructed Maashu again on how to treat the calf according to the description of the malady he described to me. Saying good-bye, I reminded them I would see them the next day at church. Again, I thought to cover the distance back in as short a period as possible. As I walked I began to think of all that the Indians had done for me over the past years and in reality, of all that they had come to mean to me. I thought of all the beauty around me as I walked along— the hanging vines, the huge trees and the odd bits and pieces of vegetation that was all so new to me. When I arrived at the river crossing below our house, I was shocked to realized I had covered the distance so quickly. I thought back over the trip and couldn't even remember the points along the way. I was thoroughly and completely happy and content—I should have been bone-tired after walking a full hour and a half, bashing through the jungle looking for a hidden calf with only a short twenty-minute rest. I was tired but the contentment of my mind seemed to totally obscure the fatigue of my body. I got in the canoe and quickly paddled to the other side, climbed up out of the canyon to our house.

What a peculiar day! I wondered what I had accomplished. Maashu could have done all that I did and never missed a beat. I had not accomplished anything at home and I was too tired to start any project now. Yet, without understanding it, I had a wonderful day.

Everyday Life

...trusting Maashu with my life...

One very beautiful warm morning, Maashu and I decided to visit the Shuar across and up the river. Visiting the Indians with cattle as we went, the trip up and back would take four or five hours. So we crossed the river below our house and tramped the up-river trail.

Late morning, we came to the farthest point up-river of our trip and were about to take the trail that forded the river at that point. I was in the lead at the moment and arriving at the water's edge, I took a look and concluded all was well. I stepped into the water and within a couple of steps was up to my thighs in powerfully moving water. Suddenly, Maashu yelled for me to go no farther.

"Don Donaldo, come back! You don't know the trail across." Indeed I didn't. More than that, there were steep rapids just 20 or 30 yards down stream that looked pretty deadly to me. I had not seen them before.

Cautiously, using my walking stick, I backed out of the water to where Maashu was standing on the beach.

"Don Donaldo, you do not know the trail across the river. There are holes and rocks you do not see. If you fall, the water will wash you down over the rapids. I know where they are. I will lead you. But first I must take both walking sticks and go out to reacquaint myself with the trail."

Taking a walking stick in each hand, he walked cautiously out to mid-stream and then backed out slowly to the beach.

Smiling, he reassured me, "Yes, now I know. I will take a stick and you take a stick. We will hold hands and I will lead you."

I thought to myself, "Now you know that you too, could have fallen and you would have gone over the rapids. I could not save you. You knew that but you checked the trail first. Maashu, you are quite a man."

So hand-in-hand, walking sticks in our down-stream hands to lean against, we started out. The water got to well above our waists and was very powerful. Step by step we made it across, a distance of maybe twenty yards. Climbing out of the river gorge, we got on the down-river trail and trudged home. After that incident I held Maashu in much higher esteem, believing he was a closer friend.

I should say that Maashu's wife, Yamainshi was a good woman and a good Christian with whom Maxine got well acquainted. The two of them became very good friends. Maxine enjoyed working with and teaching their children as well.

Until we left Macuma, Maashu and I worked together and enjoyed a close friendship. He is one of those personalities you never forget. I can thank the Lord for what He did through the two of us over the years of our friendship.

…flying calves to the new jungle communities…

Community development broadly speaking was a good-sized part of our ministry, and agricultural development was an instrument used in building communities. We encouraged the Shuar to lay out land units of 50 hectares per home and then to clear land for gardens and pastures for cattle. Ultimately and if sufficiently stable, we urged the community to then build an airstrip for the MAF planes. If they did all that, then we would provide a person in the community with at least two calves: a heifer and a bull calf. They could thereby start their herds of cattle. It would be required that the first heifer calf born would then be given to another member of the community. Then in turn the first heifer calf from that animal would be given to another member of the community, and so it went. That system was quite an incentive to begin to work for the good of the community.

I talked to Dave Osterhus about flying the calves in the Cessna 180. The next opportunity came and we set a date. I hog-tied the heifer calf very securely and we placed her in the airplane. I placed my insurance for a safe flight, a .22 caliber pistol, on the dashboard. I was ready to shoot the animal if it somehow slipped its ropes and got loose inside the plane. We waited a few minutes to let the calf settle down against her restraints and took off to meet the people waiting at the little airstrip carved from the jungle. It was a rather scary flight but other than the calf bucking and straining against the restraint of the ropes, nothing untoward happened.

As we flew, we reminded ourselves now and then of the experience of Nate Saint when he attempted to fly a pig to a distant airstrip. In mid flight, the cargo of one live pig got loose. It was just Nate, the pig and the Lord in mid-air. Nate wrote that he asked the Lord to keep the pig quiet until he could get back on the ground. The pig hardly moved a muscle until he was on the ground for which Nate thanked the Lord profusely. I did not want that to happen with a much larger animal on board. Dave and I flew a number of calves around the jungles. The one memorable incident that comes to mind is the calf we flew to Putuimi; it was memorable not for the flight but for the reception we received.

Arriving on the airstrip, twenty or thirty excited people gathered around to see this animal. We got her out of the plane, onto the ground and untied. She did not move for several minutes and the women gathered close to touch her. They began to speculate and argue back and forth as to whether it might be a strange sort of jungle deer…they had never seen a cow or a calf of any sort. Suddenly the calf came alive, jumped up midst the crowd. Women, children and men scattered howling with laughter. They put a rope around the calf's neck and tied her in the nearby pasture and she began to graze as if she had been there all her life.

…death of the Brown Swiss bull…

In the years prior to the 1970's, cattle on the farms and ranches of eastern Ecuador were a sort of native or criollo breed of cattle. The criollo had descended from cattle brought over by the Spanish with the conquest of Latin America and then crossed with any other breed brought in by enterprising agriculturists. The Catholic priests had brought in cattle to furnish milk for their schools, orphanages and other such institutions. Most later importations were of the British milk producing breeds, primarily Holstein and Guernsey. Due to poor animal nutrition and constant in-breeding, the end result was an animal that did not produce very much milk or meat. I have no idea why someone didn't think about importing one or more of the dual-purpose breeds such as the Milking Shorthorn, Brown Swiss or others available.

Castration was not practiced in the jungles mostly because screw-worms very quickly and inevitably attacked any wound found on a warm animal body. Farmers did not think it worth the effort and work to castrate and treat the wound. Therefore, over decades of inbreeding, the bulls developed into slow-maturing animals with heavy forequarters and light hindquarters. They were seldom ready for market before four or five years of age. By that time they had bred cows and the offspring were inbred and equally as poor as their parents: it was a vicious circle. The farmers were generally illiterate and so could not read any material on improving breeding practices even if it were available.

At the time we moved to Macuma to start our work, there were about twenty-five head of cattle in the area, all but maybe a half dozen of the younger animals had been trailed in as adult animals. They were the start of the herds in that area. Frank Drown had brought in one Holstein bull that had been breeding a few cows around the community, up-grading the stock to that degree. A year or eighteen months later, he brought in two Brahma bull calves and six Brahma heifers. We lost one of the heifers to disease and one of the two bulls was trailed to the Sucua Valley for use there. When the heifers reached breeding age they were bred to the one Brahma bull left in Macuma. They dropped their first calves about the time I entered the jungles. I immediately began to breed the bull to our own native stock and to that owned by the Indians in the community. When I could I bred him to heifers that were crossbreds of the Holstein. The offspring from the three-way cross were loaded with hybrid vigor giving us heifers that were very large, quick maturing cattle and they turned out to be very good mothers. I castrated every bull calf

I could find in order to hold down inbreeding of cattle in the communities. All such crossbred steers matured to butchering age in eighteen to twenty-four months and were larger and heavier by far than any native stock anywhere in the jungle.

After a year or so, I felt that using a bull of a dual-purpose breed would enhance the crossbreeding program. The Holstein bull was getting mean and dangerous to handle. I was definitely afraid that someone was sure to be hurt soon. Because Heifer Project, Inc. had been such a help in obtaining the Brahma calves for us, we applied to them for a Brown Swiss bull calf. A few months later, a Brown Swiss bull calf was flown into Macuma and I was overjoyed with the quality of the animal. I was convinced I had the very best animal available anywhere. I gave the calf the very best care of the calf. He grew to be a fine, gentle animal and so very easy to handle—typical of the breed.

At breeding age, I crossed him on one of my better first-calf heifers; she was a cross between the Holstein bull and a better native cow. The calf she produced was a superb calf. I thought I was on the way with a real breeding program.

A few weeks after I had bred the bull to my heifer he broke out of the pasture and wandered into a swampy area. He mired down in mud to his belly and could not pull himself out. I assume he got stuck sometime in the evening then fought to get out until he could no longer fight. A cowboy found him the next morning. With the help of a dozen or so men we pulled the bull from the mud but could not get him to his feet. The following morning he was dead from either pneumonia or the fact that he fought his heart out. This was one of the hardest incidents I had to deal with on the farm, it was devastating.

Along with the Brahma cattle, he certainly was one of the most outstanding animals I had ever been allowed to acquire. With the death of the bull, my plans for the improved breeding program ended. I had neither the time left in
Ecuador nor the prospects of another animal to replace him. All I could say was that it had been an interesting program and I enjoyed it immensely.

...the rope...

Daylight at 2-3 degrees south latitude of the equator lasts 12 hours, plus or minus 15 minutes. Pre-dawn light wasn't long and the evenings were short. Working through all the daylight hours available, I still didn't get it all done. My only day off seemed to be Sunday. I felt I had to become a workaholic to even come close.

There is a species of large bird in the jungles, which nest and hatch their eggs in caves high up on the sides of the Cutucu Mountains. The young nestlings develop large pouches of fatty oil in their breast area. The Indians find this oil a delicacy worth working very hard to obtain, even to the point of risking their lives. It is very difficult for them to access a cave in a vertical cliff.

Early one morning four or five young Indian men came to me and asked if they could borrow the long rope I had stored for whatever need might come up. It was very strong, about 200 feet long and a half-inch in diameter. I consented, asking them how they would use it.

The young Indian, who was more of a friend to

me than the others, explained they intended to rob the nests in the caves on the cliffs some two hours walk away. Very enthusiastically they asked me to go along with them. It all sounded so exciting, I really wanted to go but I felt that I had to finish my work before I could 'play', and there was always more work than I could possibly get done. I asked Catani to please take me the next time he went.

Catani listened to me a moment then very quietly and thoughtfully responded, "Yes, Don Donaldo, you have said all that before but you never go with us." And with that he turned and walked away. He never asked me again.

. . .of birds and rice and miracles . . .

Sunday afternoons I usually enjoyed a leisurely walk to take a look at the crops or the cattle, maybe a hundred yards or so into the jungle to stand in wonder and awe of what God had made.

One Sunday, as I contemplated a trip to Quito the next day, I walked out toward the fifteen acres of upland rice to take one last look before my flight out of the station. The crop was some thirty inches high, beautiful, waving in the breeze. It was a heavy crop. I had seeded heavy believing the soil could support it. I felt justly proud of it and looked forward to harvesting the rice in a few weeks. There would be plenty for the 75 or 80 Indian school-kids who depended on it for their daily meals.

I wandered out through the rice, looking to see how the heads were developing. To my dismay I noted the sheath around the heads were all tied together by webs made by worms. I began hurrying through the field, surely it was not all so infested. I had planted and harvested three or four crops and this had never happened before. There was no insecticide I could spray to combat the infestation. I was out in the middle of the jungle. There was nothing I could do but let nature take its course.

Sickened by the sight, I went back home telling Maxine we had lost the whole crop before it had hardly gotten started. I had also shown an Indian fellow who worked for me what was going on, indicating what the end result was surely to be.

Dave Osterhus, with MAF, flew me out early the next morning and I traveled to Quito. I took care of my business and was back home a week later. Anxious to know what really happened, I hurried out into the rice field. To my amazement the rice heads were all out, full and heavy. Hot dry weather was drying the grains and harvest-time was not far off. I did not understand.

I called my Indian friend to the field and asked him what had happened to the worms in the rice. A day after I had left a huge flock of small birds settled into the field and stayed a day or so. Once they flew on, he had checked the rice and found that all the worms were gone. The birds had made short work of them. The heads formed full and large. Now we would have rice.

I asked my friend what he called the birds. He answered, "How can I know their name when I have never seen them before?"

Never in his life, nor had anyone else for that matter, seen those birds in the jungle. Surely a miracle of the Lord, another everyday sort of miracle in the jungle. There were lots of them.

That evening, I thanked the Lord once more before going to bed.

...the forced medical furlough in Shell Mera...

In August and September of 1965 I had been working very long hard hours. I had not been out of Macuma in four or five months. Though I didn't understand how those facts could affect me physically, those circumstances did tear down my body's immune defenses. I was tired all the time. As I read both my letters and Maxine's letters to our parents written during that time, I am surprised at how much I wrote of my fatigue. I began to feel that something was wrong with my urinary system so went out to see the doctor at the HCJB hospital in Shell Mera.

The doctor prescribed a medication that did not seem to help over several weeks. Returning to him, he decided the problem might be in my kidneys. The first week of November, in a procedure at the hospital, he inserted dye in the kidneys and then took x-rays of them. That evening while lying in the hospital bed I began to feel intense pain in my back over the area of my kidneys. I lay there trying to endure the pain, thinking that in a minute or two it would go away but it did not. I called the nurse and she in turn called the doctor. He determined that the dye had caused my kidneys to shut down. We did not have kidney dialysis and if the kidneys failed to start working I would die. He injected a drug to stimulate the kidneys. Then he told the nurse that was all he could do and that he was going upstairs to his apartment to pray. I knew then it was serious, he was not kidding! The doctor later told me that he asked

God to spare my life, reminding the Almighty who knows all that I had two young children to raise, one of which was handicapped. After thirty minutes passed, I needed to urinate. The nurse ran to tell the doctor who determined that my kidneys had restarted and were functioning. We praised the Lord for His kindness and mercy to me.

The next morning the doctor came in to my room for a serious talk. I sensed that he was not there to cheer me up. In a very stern and no-nonsense voice, he informed me that my body was very fatigued, that I had abused my body by working too long and too hard. Now I was going to have to pay the price. I was not to return to Macuma for a month and if I had a problem with that he would talk to the GMU field counsel and they would give me the orders. I lay there and cried for a while, knowing but not wanting to admit that he was right. I was flat-out tired. To be away from Macuma was almost unthinkable, there was just too much work to be done for me to stay out.

The GMU missionaries at the Bible Institute talked to Maxine on the radio and she came out prepared to stay the month. We made arrangement to take up a room at the Institute. The price I paid meant spending one boring month in Shell Mera doing nothing but reading and resting. Maxine accepted the same fate as she would need to care for me.

With the month over and the written permission of the doctor in hand, we returned to Macuma and continued our work there. The night of intense pain, the uncertainty of the outcome, the healing by the Lord and the boredom of resting, brought me to deal with myself in a much more intelligent manner. It was an experience

I have not forgotten and don't want to repeat.

> *...a baby without a name is hardly a person...*
> *and if not a person who should care...*

Early November, a Saturday morning and the sun was high in the sky. The wet season had really been wet. Finally the skies cleared, giving us a beautiful day. From our porch I looked down across the little canyon toward the main trail, a jungle highway, that came up from the river crossing. We watched individuals and groups come up that trail, some came bearing sick family members. Others came in large family groups on Sunday morning for the church services. Children from across the river came, returning to school.

That Saturday morning I stood and watched the procession of visitors. Sando Sando and his young wife were walking up slowly—it was quite a climb and they moved slowly as the trail was so slippery. I watched them, realizing that Sando's wife was carrying nothing. I thought that peculiar as she had delivered a baby only a couple of weeks earlier. After a woman gives birth to a baby that child is never more than arm's reach from her—at least for the first month or two.

Later that morning, Sando came to talk about some of his cattle. Both he and his father had cattle in the same pasture and some were sick. He needed medicine and I sold him the supplies he needed. We finished the transaction and sat visiting a few minutes.

"Sando, your wife is not carrying your little son?" I asked.

He countered as usual with a question, "Is there a son to be carried? Did it not die a week ago?"

I was surprised; I had not heard of the child's death on the jungle grapevine. I said I was sorry about their loss and hoped they were not too sorrowful.

"Why should we be sorrowful?" he said. "Did someone die? It was no one, it did not have a name."

Conversation with the Shuar can be slow and thoughtful—very sensitive at times. One has time to think. I ran his question and statement through my mind again to be sure I had heard correctly. A baby had been born. The baby died unnamed. Therefore, I concluded, a baby is not a person until it has a name. If a baby doesn't have a name it is a nothing. I didn't want gringo culture to get mixed up with Shuar culture so I said no more—my question had been answered. I would ponder this and then ask someone about that concept in their culture. Why don't they give their babies names right away? At what age are they named? Why is a name so important to being a person?

We talked a few more minutes. Sando thanked me for the medicine; said good-bye and went back down the trail toward the river.

On a Sunday morning in December I sat in the church service thinking over the past few months. Thinking about the people I saw sitting there. They were singing and everyone seemed happy enough. Chirimach and his wife sat with their young baby—this was a baby born onto a piece of corrugated metal. Sando and his wife were there. Surely another pregnancy would soon be evident—at least I was sure they hoped so as babies are important to the Shuar culture. To the husband, he is man enough to sire children; to the wife, she is woman enough to bear him the children he will gloat over. There were old women paying more attention to their very small grandchildren than to the church service. An occasional dog wandered through. A very small chicken escaped the clutches of a small child and ran 'peep-peeping' across the floor.

Anguash rose to speak this morning. He was a very intelligent and perceptive young man. He was very good looking, married with a not too happy marital past. But now he was right with the Lord and desirous of preaching the Word of God to his people.

With the usual introduction he began to speak about Christmas, an event unknown to the Shuar before the missionaries came. Anguash talked about Jesus, the baby God sent to be born on this earth. He talked about the baby's mother Mary, the angels, the shepherds, and other circumstances. There was something special about this baby, he said. In fact several things were very special and he wanted to tell his people about them.

He began, "Now a long time ago, a long time before this baby was born on Christmas, the old prophets told their people that this baby was to be born. We Shuar cannot tell such things. Do we know when children are going to be born to us. No, of course not. But God, who is all knowing, knew that this child would be born. He knew because he was going to send him. He was to be his son, God's son. God wanted to prepare the people for the coming of this child so he told the old prophet to write it down in their books. They did so faithfully. The people waited and they waited and they waited until they thought God had forgotten his promise to send His son.

"Then on the day appointed the angel Gabriel came to Mary, a girl who had not slept with any man. The angel told her she was soon to have a child."

Then Anguash said something that caught my ear and I began to understand a bit more about Sando, his wife and the baby. He said, "God named the child Jesus before he was born. Do we name our children before they are born, no. He would be born, he would have a name, he would be a person and he would not die—as our babies sometimes do. We do not name our children until after we are sure they will not die as a baby. If our babies die without names, it is no one who dies, only people with names are persons."

I began to understand; an Indian baby is not named until sometime after birth for fear that it will die—as they do so many times—and if it dies with a name there is a great deal of sorrow. The name has died. Without a name it is a nobody, without a name an *it* died. But Jesus came, was born to Mary and had a name before he was born because God knew he would not die. How sad, the Shuar Indian was protecting himself or herself as much as could be done against continuing sorrow. A baby without a name is a nobody, an *it* and no one cares if an *it* has died.

...the riot in Quito at the American Building...

In the last year of our time in Ecuador, I had an interesting experience that could have turned deadly, at the least, injurious. I found myself in the center of a small riot that was protesting the presence of the U. S. in Ecuador. The rioters besieged the *American Building* in Quito, attempting to destroy it.

Before our arrival, the Peace Corps had been working in Ecuador for a few years in the rural areas of both the jungles and the mountain villages. They attempted to better the lives of the poor through better agriculture as well as public health. Individuals of the Peace Corps had ridiculed our work in the jungles. They contended that they would do their work in two years and then leave whereas we missionaries were building empires requiring our continued presence. During their stay they made some very ludicrous mistakes in common sense, bringing down on them the ridicule of the people. Those of us in the jungles were making advances that drew the accolades of the government and the people. Finally, some months before we were to leave, the USAID and the Peace Corps invited all the missionary agriculturalists, Peace Corps and USAID members working in agriculture to a conference. In essence they wanted to know what we did right and what they did wrong.

About mid-morning, we heard windows breaking. The building caretakers alerted the conference that a crowd outside had gathered and were throwing rocks at the building. We all dispersed to top-story rooms of the building to watch and decide what we were going to do. At the same instant, the steel door at the front of the building rolled down, closing with a slam and quickly locked. There was no way for the rioters to enter the building from the front.

On the top floor, I watched as rocks were thrown and windows shattered. We were told there was no way out of the building without letting the rioters in. So we waited.

I saw a large tanker truck coming down the street. I knew it was a gasoline truck because it was painted red. All such vehicles carrying flammable materials are painted red in Ecuador. The rioters stopped the truck. I thought they would take gasoline from the truck, throw it on the building and set it on fire. Fortunately I was wrong. Using a fire-axe from the truck, they hammered at the steel door, trying to break through into the building. Some of the group in the building had begun to look for a way out. They found an exit through the basement, out a door into the backyard and then over the brick wall. Like most walls in Ecuador, the top of this brick wall had pieces of broken glass embedded to inhibit thieves climbing over the wall. We found a short ladder, the glass was easily broken off and people began their escape over the wall.

I suggested to a friend with whom I had come that we go around to the front of the building and see what was going on. The friend was very fearful and wanted no part of the plan. The next morning's newspaper reported the riot as not a big deal. It was an interesting experience.

> *...the hydroelectric plant*
>   *and the burning of the Hercules C130...*

Upon our return from furlough to Ecuador in 1967, Frank and Marie Drown and their family left for a few months vacation in the States. We took the reins of the station along with the help of Yoders, Sara Watkins and one or two others.

After several months the Drowns returned to the work in Macuma. While on furlough, Frank had been given a hydroelectric plant that he thought could be used in the jungles. Before returning, he took action to dismantle the plant and ship it to Ecuador.

Knowing we would have to leave in two or three years, I wanted to finish all the projects I had started, to put in place all things that would make the work easier for any replacement that might come to Macuma. The hydroelectric plant was not part of my original plan.

Though this was not part of my plan, it became obvious that now I would have to become involved with preparing for the hydro-electric plant. We started out by asking for the services of a surveyor from HCJB, Gordon Wolfram who was a licensed Civil Engineer. He came to the jungles and did all the surveying for the proposed canals and plant placement. After working for two weeks, he left us with an excellent layout plan from which to work.

We made plans and preparations for the arrival of the plant via Alaskan Airlines' Hercules C130 cargo plane. We lengthened the airstrip and built storage space for the parts that needed to be covered. We built a road to the place along the river where the plant would be located. I find that the times in my life when the level of risk to life, limb, financial well-being and mental credibility is high leaves me mentally uneasy with apprehensions. Such apprehension coupled with anxieties can possibly bring on premonitions relating to the situation I am working in. The anxiety has a way of managing my activities, providing priority to the sorts of risks I allow myself.

This was true as we worked and lived in Macuma. All reasonable travel to and from Macuma was by airplane. While we could walk out, such a trip at that time was hugely impractical.

One thought seemed permanently engraved in the back of my mind influencing everything I did or thought to do. As we planned for the arrival of the Hercules C130, we had better be correct in all we did the first time because there was unlikely to be a second chance. For me, this became a primary and unwritten consideration. In other words, throughout our daily lives, we understood that no one had better get hurt or sick after four o'clock in the afternoon because the airplane could not fly in and back out before darkness fell at about six o'clock. Any emergency late in the day would have to be taken care of on-site or wait until the first possible radio contact at seven the next morning. Ultimately, in an irreversible situation, there was no way to get outside help to right something gone terribly wrong.

With that consideration in mind, I seemed to be overwhelmed with a foreboding anxiety concerning the impending flight of the Hercules C-130 cargo plane. The thought that something would go wrong and prevent the plane from leaving our airstrip was a constant nightmare for me. My anxiety as we discussed landing the big plane on our grassy strip was gut-wrenching. I did not feel secure in our preparations. There always seemed to be something else that maybe should have been done to make the situation more comfortable.

Even in anticipating the arrival of a DC-3, a much smaller plane from the Transportes Aereos Oriente (TAO), left me in a deep agitation. Their pilot was the best in the business. The plane could easily operate out of our marginal strip. But I did not relax until it had been unloaded and safely taken off. Only after they were safely in the air could I walk away without looking back.

Any number of factors under my responsibility

could become a danger to the safe operation of the large planes. I worked day and night to remove or reduce any possibility that something would happen to endanger their landing and take-off. The magnitude of my concern grew exponentially as we began to prepare for the arrival of the big Hercules C-130.

We had lengthened the airstrip to one thousand meters as required for this large craft. We improved the air strip as best we knew how. Even in the best conditions at its driest, the strip was marginal for the large plane. Seeking a clear conscience, I never fully agreed that such a plane could land on it with reasonable safety.

There were two flights planned bringing in a total of fifteen to eighteen tons of equipment for the projected hydroelectric plant. Frank Drown made contact with the company handling all the work involving contracts and loading the equipment in Shell Mera. He was very good at such work.

After the agreement was reached, my responsibility began in providing a safe environment for the landing, unloading and getting the plane back into the air. The only thing I was sure of was the ability of the plane and its pilot. Missionary Aviation Fellowship flew their pilot into Macuma. He looked around, pushed a quarter-inch rod into the ground to make a judgment as to whether or not the strip was solid enough to sustain the weight of the plane. He judged the surface ready and returned to bring in the big plane. The first flight arrived in the early afternoon of a hot, dry day. I don't know when but somewhere prior to the actual landing, the principles of Murphy's Law began to take affect.

The landing itself was uneventful, yet, as they rolled to a stop the wheels on one side sank into the ground about fifteen inches. I led a group of men unloading the cargo while Frank and some of the Indians began digging a trench and lining it with rock. We were attempting to make a more solid base to support the weight of the plane. The pilots would make their engine run-up while sitting in the trench. Then, under full power, they would make their take-off run from that point, hoping the rocks would sustain the great weight. They hoped that the lift afforded by even a slow speed would prevent the plane sinking again. The distance from that point to the end of the airstrip was five hundred meters—theoretically and officially insufficient to make a safe take-off.

The pilot and crew climbed into the plane and went through their pre-takeoff and engine run-up procedures. With all systems signaling a go, they waved good-bye and focused their attention on the runway ahead. Along the air strip, we focused our attention on the landing gear and the ground under it. I was gripped by fear and gut-wrenching apprehension as the pilot applied power and the plane began to roll forward.

Every inch forward was a triumph. Anticipation of the next inch was sheer terror for fear the wheel would sink again. The plane rolled out of the trench and the soil held the awesome weight. The plane picked up speed and lift. The pilot held the wheels to the ground passing the point of no return, that point beyond which he could not stop before reaching the end of the runway and the canyon beyond. At the last possible moment, the pilot lifted the big plane off the ground, soaring into the air toward the Cutucu Mountains beyond.

I ran down the runway to see exactly where he had lifted off. I wanted to know the exact distance he used to get airborne. A slight rise in the center of the airstrip prevented us from seeing the ground at the other end of the strip. As I reached the lift-off spot I looked toward the mountains and I saw that the fence at the end of the runway was down. The plane's under-carriage had knocked the fence down as the wheels left the ground, passing over it. Had the pilot kept the plane on the ground another ten feet, I felt sure it would have been catastrophic. I never decided whether providence or skill got the plane into the air that day. Probably it was the providence of sufficient skill for and at the moment it was needed. The success of that trip encouraged the immediate preparation and planning the next flight. I shuddered in fear that we were pushing our capabilities. One success did not by any stretch of the imagination insure success for the next landing. I felt the margin for error was too great.

Once again we waited for dry weather, needing at least a week of hot dry weather to dry out the soil on the strip. A week without rain was unusual except in times of temporary and unusual drought was almost more than we could hope for. When the dry weather looked as if it would hold another two or three days, Frank Drown again flew out to make preparations for the next flight. Again, we did all that we knew to do to prepare for the next and last landing. As before, all livestock was moved to pastures far from the airstrip. All the necessary equipment was standing by and prepared for unloading the C130. We called in as many Indians as possible to help with the operation. Every arrangement was made on the premise the weather would hold. And it did. Again, the day arrived.

Economics dictated that this trip originate from Quito, so the Herc was loaded there and flew directly to Macuma. The plane left Quito, climbed up and over the eastern cordillera of the Andes before beginning the long descent out over the jungles to Macuma. The view must have been magnificent as it was a beautiful clear day.

I told the Indians gathered for the event that the plane had taken off and they initiated a game of who-first-to-sight-the-plane. A young boy claimed that honor. There were no clouds to impede the pilot's sight. No wind on the strip as the plane came in very low over the trees. As I stood and watched, I begged and pled with the Lord. I cried out in an agony I have seldom known in my life, pleading for a miracle. God saw things otherwise than I saw them. I did not then, nor have I since been able to imagine why the accident happened without forewarning.

The plane dropped on to the ground at the eastern extreme of the strip and braked to a slow stop beyond the exact middle of the strip where we had instructed the pilot to stop. I was afraid. As if we had practiced a dozen times, the plane was unloaded quickly and without mishap. All equipment and cargo was moved to the side of the strip out of the operating area. I was proud of the Indians who helped, laughing and joking as they worked.

Along with the pilots, we had planned for the airplane to turn, and while in place to make the engine run-up. The pilots would only then accelerate for the lift-off from where the plane stood, taking off to the east. With no wind and six to seven hundred meters of runway, they could make it.

We cleared the hundreds of Indians off the runway. The pilot and crew climbed aboard, shut all doors, started the engines and began their run-up. As the engines reached their required revolutions the ground under the wheels could stand no more and gave way to the great weight. The left side of the plane began to sink. As it did so, the left wing dropped and the tips of the blades of the out-board engine propeller struck the ground sufficiently to break loose, walking over to slam into the in-board engine. That engine erupted into flames. The flight engineer applied all the on-board fire extinguishers available but they were not enough. The flames grew quickly beyond any hope of extinguishing them with what was available. My nightmares of previous nights came alive, prophetic after all.

For a few seconds we stood stunned. The pilot quickly shut down the other two engines and the six men in the crew dropped out of the emergency exit asking what fire-fighting equipment might be available. Surely they knew there would be none available out here at the end of the world. Suddenly, they too understood the finality that there was no second chance out here. The earlier flight had not been an indication that there would be success with this flight. Over the next thirty minutes the fire gutted the fuselage of the plane. We were left with a burned out hulk, both wings drooping with the outer ends resting on the ground.

When the fire started I did not know what to expect if and when the flames reached the fuel tanks. I asked the crew and none of them could give me a sure answer. For all we knew, exploding fuel tanks would blow the plane to small bits, leaving a crater in the strip. I did not know what to expect from the fire. I did know that there were several hundred Indians within one hundred feet of the plane who could be killed or maimed if such a blast did occur. I had to get them as far away from the danger as I could.

In both the Spanish and Shuar language, I yelled and screamed at them to run but to no avail. The crowd would not move. Men, women and children of all ages milled around watching the fire devour the plane, not knowing what to do. One man just stood and laughed at me. In desperation, I suddenly lashed out at him, hitting him so hard with my fist I knocked him down. I screamed at him to run, to get away. My actions got the attention of the Indians. They started running and within seconds the airstrip was vacant except for the burning Hercules.

I ran to a point one hundred-fifty yards away, then stood and watched. As I watched the plane burn, I felt so alone. Frank Drown had gone to the radio to ask for any assistance available. MAF flew in with all available fire extinguisher but it was far too little and too late. The pilots and crew of the Hercules stayed overnight with us. The next morning MAF brought their two planes in and picked up the crew, flying them out to Shell Mera where they could hire a taxi for the trip to Quito. There was nothing more they could do in Macuma.

The crew felt very sad that things happened as they did. To them it was just an accident they could walk away from and pick up the next task assigned them. The plane was insured for more than its purchase price. A few weeks later we received a check for some $200. from the pilot and crew; a gift from them to us on the station.

As I related above, the Indians were not in the least interested in running from the burning plane and did so only as I reacted violently. One humorous story I heard was of a little boy who, when he saw me knock the man down decided it was time for him to run. He ran down to the river, crossed it and ran on into the jungles. He knew of a hollow tree, climbed into it and stayed there for two days. His parents were frantic at his disappearance not knowing what had happened to him. He said that after two days and never hearing a big boom, he figured maybe it was time to go home. His arrival brought great relief to his frantic parents.

I had no way of knowing the hours, days and weeks of work I would have to put in to clear the wreckage from the airstrip. After the four engines and odd parts were removed from the hulk, the rest of it was nothing but non-biodegradable trash. An environmental hazard in the jungle that had to be put somewhere out of the way. The jungle growth would cover it in time but the wreckage remained. I hoped the acidic jungle soil would destroy it more quickly than we expected. We did salvage and sell several hundred pounds of lead and used several hundred gallons of kerosene fuel. The fuel eventually caused a great deal of damage to the diesel engines where it was used. We salvaged a few hand-tools and the Indians used some of the sheet aluminum on their houses. The work savaged me in a battle with bitterness and depression.

The salvage work had nothing to do with the agricultural development I was there to accomplish. I was required to keep the airstrip in a safe operable condition. I felt as if the crash had been due to factors beyond my control. Compared to the work with the

Shuar and the cattle, the time required for the salvage seemed so senseless. As we tore apart the plane and removed the wreckage from the strip, each day added to my bitterness.

The hydroelectric plant did get built and was used for over thirty years. I have no way of knowing whether it was really worth the financial costs to all concerned. Was it worth the cost to my emotions and my physical body. What did it cost the community of Indians? I really can't agree that the costs in labor and finances were worth the benefits received. Only the Lord knows and I believe this is one of the secret things that God withholds from us as we are told in
Dueteronomy 29:29.

Two weeks after the incident, Alaska Airlines sent a man into Macuma to salvage all that was worth salvaging from the plane. With that authority, he salvaged the four engines, one of which was a burned out hulk, some electric motors and pumps from the fuel tanks in the wings.

His living quarters were a small building with cooking and bath facilities. Two missionary couples shared the responsibility of feeding him lunch and supper. After a period of ten or so days, the man became restless and uneasy with his life there in the jungle. He had never in his life, much less his wildest imaginations, been in such a place so far from civilization. The only social life available was conversation with the two missionary couples and that got old for both parties.

He asked for a plane to come in and take him to Quito for a few days vacation. When he returned he brought liquor with him, which we had refused permission to bring in. There was a world of difference between his life-style and the only one that existed in Macuma. The man attended our Sunday church services including the Sunday evening worship services with the Spanish school teachers on the station.

Frank Drown talked a lot with him in the evenings explaining our beliefs and the Bible. He was one of the few gringos we had met who knew nothing whatsoever about the Gospel of Jesus Christ. Shortly before he finished his work and left, he did confess to acceptance of Jesus Christ as Savior. I have no idea what his life was like afterward. We never heard from him again.

...the last leg of the journey...

After the disastrous accident with the plane, the agenda for using my time changed drastically. We had only so much time left in Ecuador and the most pressing work was to get the remains of the C-130 salvaged and off the airstrip. All else took second place.

After four or five months the job was complete and I returned to completing the agricultural development projects for which Macuma was responsible. I did not want to leave projects unfinished, hanging over the other missionaries. After evaluating what remained to be done, we decided that the installation of the power plant would be delayed and left to someone else after our departure.

In our last few months, Jim and Norma Hedlund arrived to begin work with the radio ministry. The facilities for the radio station were enlarged. After we left, Jim became heavily involved with the hydro plant, helping to design the electrical system that would control the plant. Over the next forty years, Jim and Norma worked with Shuar translators on a fresh translation of the Shuar Bible. The day came when they could no longer wait to publish the new translation though seven books remained untranslated. The celebration brought the community together as the elders presented the newly printed Bibles to their people. Maashu, the man who worked so closely with me, had fallen away in the years after I left. He was first in line to purchase a Bible.

In time, he confronted the Hedlunds claiming to have found errors in their translation. They politely smiled and encouraged him to keep reading, checking their accuracy. The Spirit convicted him of his sins and Maashu returned to his faith. His wife, Yamainchi, was struggling with health issues and he began to care for her in a way that demonstrated a new life blooming within him once again.

Along with the expansion of the radio ministry, the school facilities were enhanced to be used more efficiently.

The Alliance Academy requested a conference that year and informed us that they felt they were no longer able to teach Garth. At that time, hiring a special education teacher was not an option. They requested that we return home so that Garth might be educated further to improve his opportunities as he grew into an adult. With heavy hearts we knew the time had come to leave Ecuador. We had dreaded this moment from the day we first landed in Guayaquil.

I dismantled my end of the organizational structure, selling all but five or six of the livestock on the farm. The pastures would be allowed to return to jungle growth. The barns were dismantled and the lumber used to build other facilities. I settled any and all projects with the communities with which I was working. There would in all probability never be anyone to replace us. We had completed our work in Ecuador.

CHAPTER SEVEN

A TIME TO RETURN HOME

There is no welcome committee waiting at the gates as a missionary returns home one last time. There are no companies anxiously waiting to bid on the services of a missionary that has spent years overseas. If anything, as the missionary attempts to pick up the pieces of a life that they left long before, they are met with a puzzled attitude.

"And why did you do this?"

There are no neat little spaces to explain missionary services on employment applications. Coming home feels as if one has mis-stepped and lost one's way.

The only thing easy about returning to the U S from Ecuador was the mechanics of the move. We did not linger and prolong the sadness of the event. We bid farewell to those for whom we cared much. In order to cut shipping costs, we disposed of as much of our household goods as we could afford, then packed and shipped home a couple of barrels and several boxes of our belongings. With the date of departure set, the emotional trauma of saying good-bye to Ecuadorian friends, missionaries and a wonderful lifestyle was so difficult.

Knowing we would not be able to make the trip later at a more leisurely pace, we decided to take the long way home and see as much as possible on the way. The long way home would give us time to say goodbye to the country we had left and to prepare for a new life. We flew from Quito to Panama to San Jose, Costa Rica and on to Guatemala City, Guatemala, stopping for a day or two each place. After crossing the border into Mexico we became tourists visiting archeological sites and museums.

Arriving in the United States, we could not afford to dwell on the sadness that permeated each of us. We had to get on with our lives. My career changed. I interviewed with corporations and companies involved in agriculture from San Francisco to Fresno to Phoenix and was turned down because of my seven years in missions in Ecuador. Most interviewers put me in the same category with the Peace Corps, missionaries and other non-profit organizations. Their perception was that I would not be profit-oriented.

Failing to acquire a job in agriculture at the level at which I thought I was capable, I could wait no longer; I had to take a job that paid a living wage. From those high and lofty aspirations I got a job in Phoenix, as a welder/mechanic at a Ford Tractor agency at minimum wage— for one dollar and five cents an hour—for a 45-hour work week.

Summertime is always hot in the Phoenix area and the summer of 1969 was no exception. Our return to Glendale and the old community proved again that after leaving one can never 'return home.' What we left had disappeared as former friends had moved on. We had to rebuild our home and our lives.

Our time in Ecuador was now relegated to the past, to be returned to only through pictures and memories. We continued our commitment to the Lord and His direction in our lives, to His direction by providence or that most-times inaudible, 'still small voice.'

Six to eight months after we settled in Phoenix, we realized that we must have two vehicles. I needed a vehicle to get to work and back every day. Maxine needed a car for errands, meeting appointments for Garth and to meet substitute-teaching commitments. Early in 1970 I bought my first motorcycle. It was a Suzuki Hustler X6 250cc, two-cycle engine street-bike, reputed to be the hottest bike of its class at the time. I think I paid $250 dollars for it in something of a damaged condition.

After getting the bike repaired and back into a legal street-ready condition, I received my first traffic citation the first time out. I intended going somewhere on an errand and got six or eight blocks away when the engine died. I called Maxine to bring some item from home so I could fix the bike. With it fixed sufficiently to ride back home, I quickly threw my helmet into the car, started the bike and followed Maxine home, a distance of maybe six blocks. I had no taillight as I rode through the late afternoon's golden glow. We went through the residential area to stay off main streets. Would you believe a policeman spotted me as we crossed an intersection? Only two blocks from home, he gave me a ticket for no taillight and not wearing my helmet!

Though it was a street bike I used the little bike for rides into the desert via dirt roads, a means of getting out alone to relieve some of the stress of working hard in the heat. The bike took a beating to say the least. Gasoline was about 15 cents a gallon and it cost 67 cents to fill the tank - a cheap commute to and from work. The two-cycle engine was hard on spark plugs so I had to replace them often. I remember it taking two weeks saving our extra change to buy one spark plug for $2.50.

I married for love and companionship so my next goal was to get Maxine on the bike, hoping to gain her approval for future trips. This was no small task. One late afternoon I borrowed a helmet for her and convinced her to take a ride with me. She insisted on a trial run around the block. I knew such a short distance was not enough and so just kept going for some five miles. Arriving home she was so angry with me that she had nothing to say to me for three days. Well, maybe she did have something to say to me but figured if she said it, it would do nothing good for the health of our marriage. Little by little she became more accustomed to the bike.

In the years that followed larger, heavier motorcycles, worthy of long trips, would follow as we crossed the western United States. We bought a small trailer after I retired and enjoyed seeing the country.

Toward the end of 1974, we received a letter from Joe Sanders, Director of Message of Life, Inc. in the Sierra of California. He asked if I would be interested in joining their organization supported on a faith-basis as a missionary. I had known the Sanders family since I was a young boy. The ministry was located just south of Yosemite National Park. I was eager to make the move.

After six months of writing letters, speaking and praying, we had gained $450 dollars a month in support. We moved into a very small mobile home on Ministry property. Maxine and I went to work in the darkroom and Maxine in the assembly area. The experience I gained in working in the darkroom was invaluable.

The time came when we could no longer continue working at Message of Life and chose to return to Santa Barbara to help care for our son and for my parents. Leaving was difficult as our lifestyle in the Sierra foothills was my ideal. I hoped we could stay there the rest of our lives as it was a wonderful place to live and to work.

Once again, we witnessed the graciousness of the Lord in our move from Ahwahnee to Santa Barbara. First I contacted a rental agency and they directed me to a house in the Santa Barbara Shores district, to the west of Goleta about as far west as the city suburbs extend.

I was moving into the area with no firm job. The man who owned the house was a pleasant man but always pessimistic about any present situation. He was cleaning up and repairing one of his many rental units that had been badly damaged by the previous renters. He said he had a house that was to be open in a few days on another street; the price would be $435 a month.

I chose honesty as the best approach and told him what I was looking for, that I had no job, only a firm interview. I had enough money to pay for two months rent. I said that as a Christian I would be honest with him and I would maintain the house in it's present condition. I told him a little about Garth and that the reason we were moving to Santa Barbara was to get him settled. He had a son in a similar condition. He accepted my story and I paid him the first month's rent on the spot. I had the house, now the job. I made a list of all the print shops in Santa Barbara, relying on the experience I had gained at Message of Life, once again witnessing God's provision for us. Over the next few years, I worked for three different printers, At various times, the men I worked for were good employers, yet

I struggled between the desire to perform well while working for men who were not concerned about their employees.

We planned for a two-year stay in Santa Barbara intending to settle Garth and then move back to a rural environment. After two years, it was evident that Garth would need a *counsel and care* presence nearby for the rest of his life and as his parents, the responsibility fell to us.

In 1988, when we learned our rental home was to be sold, we were able to buy an older mobile home. We found it to be airy, warm and comfortable. In June 1989, just a few months after purchasing the mobile home, Maxine collapsed with a broken aneurysm. Over the next few months as she struggled to recover there were many times when I asked God why he had allowed this to happen after a lifetime of conscientously working, saving and loving each other.

It is hard for me to believe that anything in the life of the Christian happens without a reason. I *don't* believe anything happens without the Lord knowing it. Sometimes it is hard to believe and act on that belief, nevertheless I do. I make choices: I don't try to second-guess God.

In time Maxine recovered, learning to talk and care for herself again. She retired from the stress of working. A few years later I retired and we continued to travel. In 1999, I sold my motorcycle and we made a few more trips by car. Most of our time was spent quietly in Santa Barbara.

In 1986, Maxine had been diagnosed with Amyloidosis, the type that invades major organs and as the years passed she grew weaker. I also struggled with my health. I supposed it did not help that I fell off a ladder once, puncturing a lung. And then I fell off a horse, just once, when riding with my sister. By October 2013, I had been in the hospital twice with pneumonia. Each time, Ruth had rushed over to help with my care and to assist Maxine with the details of life.

In 2004, Garth married a woman he had met in the church orchestra. Five years later, he collapsed from a major stroke. In the tests following the stroke, the doctors learned that one side of Garth's brain had never developed causing the cerebral palsy. Yet, considering all he had accomplished over a lifetime, he had been blessed. His wife allowed him to pass on to the God he loved throughout his life.

Ruth and Ken's family grew to include a son and two daughters. In 2005, they lost their son in the conflict in Iraq. Marty's death shook me to the core.

In November 2013, Maxine did not feel well one evening. I tucked her into bed and as I walked around the bed, I tripped and fell, engulfed by horrible pain. Arriving at the Emergency department, we learned I had broken a hip. In time, the femur below that hip would also separate. While I was still in the hospital, Maxine's heart began to fail with her heart rate dropping to 25 beats per minute. The doctor installed a pace maker. Further tests revealed a large tumor on one ovary. We understood that she would not survive much longer.

With Ruth's encouragement, we moved closer to her. Ten days after the move, I fell once again, breaking my other hip. The following day, Ruth took Maxine to the Emergency Department as her body was failing. She was placed in hospice care and died in March of 2014.

Over the next year, I underwent two hip replacements, two cataract surgeries and one hand surgery. My family suggests that I am slowly becoming an artificial man, something I never intended. I do not know why God has prolonged my life. I am 88 years of age. Yet, as I have said repeatedly, I choose to trust him for the days he has given me. In all things God has proven to be faithful.

I now live in a small apartment near Ruth and her husband. She insists that I learn to cook for the first time in my life! I am 88 years old! What is she thinking? Now, I have a crock pot.

The mountains stand tall over the town of Flagstaff. The snow comes in the winter with the winds whipping above the rim of the canyon below my apartment. I need a walk and the sun is shining. And so, I take up a cane and walk out the door.

www.ingramcontent.com/pod-product-compliance
Lightning Source LLC
Chambersburg PA
CBHW051944290426
44110CB00015B/2101